LIST OF THOSE ATTENDING NUMERICAL TAXONOMY COLLOQUIUM

2nd–4th September, 1968

Mr H. D. Abramson	Computing Laboratory, The Mathematical Institute, University of St. Andrews, St. Andrews, Fife.
Professor J. Anderson	Department of Medicine, King's College Hospital Medical School, Denmark Hill, London, S.E.5.
Mr John Barrs	Botany Department, The University, Southampton, SO9 5NH.
Mr F. A. Bisby	Department of Forestry, University of Oxford, Oxford.
Dr A. S. Boughey	University of California, Irvine, California, United States of America.
Dr A. J. Boyce	Department of Biological Sciences, University of Surrey, Falcon Road, London, S.W.11.
Miss M. Buckley	Arthritis Rheumatism Council Field Unit, Clinical Sciences Building, York Place, Manchester 13, M13 OJJ.
Mr R. M. Campbell	Computing Laboratory, The Mathematical Institute, University of St. Andrews, St. Andrews, Fife.
Professor Wilfrid I. Card	University of Glasgow, Department of Medicine in Relation to Mathematics and Computing, 6 University Gardens, Glasgow, W.2.
Mr Andrew D. Cliff	Department of Geography, University of Bristol.
Dr A. J. Cole	Computing Laboratory, The Mathematical Institute, University of St. Andrews, St. Andrews, Fife.

Professor Y. Craig	Department of Geology, University of Edinburgh.
Dr R. Crawford	Department of Botany, University of St. Andrews, St. Andrews, Fife.
Mr H. E. Cuanalo	Department of Agriculture, University of Oxford.
Mr A. J. T. Davie	Computing Laboratory, The Mathematical Institute, University of St. Andrews, St. Andrews, Fife.
Dr A. H. Dawson	Department of Geography, University of St. Andrews, St. Andrews, Fife.
Dr P. E. Gibbs	Department of Botany, University of St. Andrews, St. Andrews, Fife.
Miss Ruth Goldberg	M.R.C., Clinical and Population Cytogenetics Research Unit, Derbyshire House, St. Chad's Street, London, W.C.1.
Dr Friedhelm Goronzy	c/o Arthur Anderson & Co., St. Alphage House, 2 Fore Street, London, E.C.2.
Mr J. J. D. Greenwood	Department of Zoology, University of Dundee, Dundee, Angus.
Mr Peter N. Grimshaw	7 Walton Street, Atherton, Manchester.
Dr A. V. Hall	Bolus Herbarium, University of Cape Town, Rondebosch, C.P., South Africa.
Dr Keith Hope	Scottish Home and Health Department, Research and Intelligence Unit, 79 Lauriston Place, Edinburgh 3.
Monsieur J. M. Hubac	University of Paris, Laboratoire de Biologie Vegetale, Batiment 490, Orsay, (Esonne), France.
Dr R. B. Ivimey-Cook	Hatherly Laboratories, Prince of Wales Road, Exeter.
Mr D. M. Jackson	Cambridge University Mathematical Laboratory, Corn Exchange Street, Cambridge.
Dr J. N. R. Jeffers	The Nature Conservancy, Merlewood Research Station, Grange-over-Sands, Lancashire.

NUMERICAL TAXONOMY

NUMERICAL TAXONOMY

Proceedings of the Colloquium in Numerical Taxonomy held in the University of St. Andrews, September 1968

Edited by

A. J. COLE

Computing Laboratory, University of St. Andrews
St. Andrews, Fife, Scotland

1969

ACADEMIC PRESS London and New York

ACADEMIC PRESS INC. (LONDON) LTD
Berkeley Square House
Berkeley Square
London, W1X 6BA

U.S. Edition published by
ACADEMIC PRESS INC.
111 Fifth Avenue
New York, New York 10003

Library of Congress Catalog Card Number: 71-92406

S.B.N.: 12-179650-7

PRINTED IN GREAT BRITAIN BY BUTLER & TANNER LTD
FROME AND LONDON

Mr A. Kadwa	Computing Laboratory, The Mathematical Institute, University of St. Andrews, St. Andrews, Fife.
Mr B. K. Kelly	M.R.C., Computer Services Centre, Derbyshire House, St. Chad's Street, London, W.C.1.
Mr P. Kent	Atlas Computer Laboratory, Chilton, Didcot, Berks.
Monsieur P. C. Lerman	Centre de Calcul, Maison des Sciences de l'Homme, 13 Cite de Pusy, Paris 17e, France.
Monsieur H. F. Leroy	Departement Geologique Central, ELF.RE, 7 Rue Nelaton, Paris XVe, France.
Mr J. W. McInnes	The Weir Group Limited, Edinburgh House, Princes Square, East Kilbride.
Dr J. McNeill	The Hartley Botanical Laboratories, The University, P.O. Box 147, Liverpool.
Miss W. Maddren	Computing Laboratory, The University of St. Andrews, St. Andrews, Fife.
Mr R. L. Middleton	Edinburgh Regional Computing Centre, 4 Buccleuch Place, Edinburgh 8.
Mr D. T. Muxworthy	Edinburgh Regional Computing Centre, 4 Buccleuch Place, Edinburgh 8.
Dr R. J. Ord-Smith	Director, The Computing Laboratory, The University of Bradford, Bradford 7.
Dr L. Orloci	Department of Botany, University of Western Ontario, London, Ontario, Canada.
Dr Charles C. Ostrander	Georgia Institute of Technology, Engineering Experiment Station, Atlanta, Georgia, U.S.A.
Dr A. F. Parker-Rhodes	20 Millington Road, Cambridge.
Dr J. M. Parks	Director, Marine Science Center, Lehigh University, Bethlehem, PA. 18015, U.S.A.

Dr Keith Paton
M.R.C., Clinical and Population Cytogenetics Research Unit, Derbyshire House, St. Chad's Street, London, W.C.1.

Mr A. J. Playle
University of Lancaster, The Computer Laboratory, Bailrigg, Lancaster.

Dr D. C. D. Pocock
Department of Geography, The University, Dundee, Angus.

Mr W. A. Read
Institute of Geological Sciences, 19 Grange Terrace, Edinburgh 9.

Dr Adrian J. Richards
Botany School, South Parks Road, Oxford.

Mr G. J. S. Ross
Statistics Department, Rothamsted Experimental Station, Harpenden, Herts.

Monsieur M. Roux
University of Paris, Laboratoire de Biologie Vegetale, Batiment 490, ORSAY, (Esonne), France.

Mr Michael J. Sackin
The M.R.C. Unit, The University of Leicester, University Road, Leicester.

Mr C. P. Saksena
Computing Laboratory, The Mathematical Institute, University of St. Andrews, St. Andrews, Fife.

Mr H. R. Sanders
Ministry of Technology, Torry Research Station, (P.O. Box 35), 135 Abbey Road, Aberdeen, AB9 8DG.

Dr James M. Shewan
Microbiology Department, Torry Research Station, (P.O. Box 35), 135 Abbey Road, Aberdeen, AB9 8DG.

Dr P. H. A. Sneath
The M.R.C. Unit, The University of Leicester, University Road, Leicester.

Dr D. H. N. Spence
Department of Botany, University of St. Andrews, St. Andrews, Fife.

Miss E. A. Strevens
Arthritis Rheumatism Council Field Unit, Clinical Sciences Building, York Place, Manchester 13, M13 OJJ.

Dr T. R. Taylor | University of Glasgow, Department of Medicine in Relation to Mathematics and Computing, 6 University Gardens, Glasgow, W.2.

Monsieur R. Tomassone | Ministere de l'Agriculture, C.N.R.F., 14 Rue Girardet, Nancy, (M. et M.), France.

Professor K. Walton | Geology Department, University of St. Andrews, St. Andrews, Fife.

Mr Stephen D. Ward | Department of Botany, University of Aberdeen, St. Machar Drive, Old Aberdeen, AB9 2UD.

Mr A. J. Willmott | Department of Computation, University of York, Heslington, York.

Mr David Wishart | Computing Laboratory, University of St. Andrews, St. Andrews, Fife.

PREFACE

The original intention of this Colloquium was to enable a few local enthusiasts to get together to discuss their successes and failures in the development and application of methods in Numerical Taxonomy. It grew rapidly beyond our parochial bounds into an international meeting with representatives from Canada, the United States, Mexico, South Africa, France and Germany as well as from many parts of the United Kingdom. We, nevertheless, tried to keep to our original intention to make this a working Colloquium during which failures and criticisms as well as successes would be discussed both during and after lectures.

In all, eighteen papers were offered and these form the basis of these Proceedings. They cover both the theory and application of methods of Numerical Taxonomy and to avoid any suggestion of ordering according to importance they have been ordered alphabetically according to the author's surname or, where there are several authors, according to the surname of the first mentioned. Any discussion which followed these papers has been recorded by volunteer scribes and immediately follows the corresponding paper. The questions and comments were handed in in writing by the participants immediately after the lectures. Occasionally the question as written does not exactly coincide with that asked and this accounts for an occasional discrepancy between question and answer.

During the last discussion of the Colloquium it was suggested that a list of computer programs used by the participants would make a useful addition to the Proceedings. Some brief descriptions of such programs listed under the name of the contributor are given in an appendix. Further details may be obtained from the appropriate contributor.

My thanks are due to many people who helped to make this Colloquium a success. In particular I would mention Miss Walker, the Domestic Bursar of Southgait Hall who fed us so well, the Chairmen of the various sessions who kept us in order, the volunteer scribes who recorded the discussions and, of course, the speakers themselves. Professor Sokal of the University of Kansas who was unable to attend the Colloquium because of other engagements kindly wrote the Foreword to these Proceedings. Finally I must thank my secretary, Miss Elizabeth Crawford who typed the first draft of these Proceedings at the same time as performing her normal departmental duties and also Mrs Connie MacArthur who typed the printer's copy. Any credit for the presentation is theirs and any blame for undetected errors is mine.

A. J. Cole
Editor

CONTENTS

FOREWORD

Numerical Taxonomy seems to have come of age. The signs attesting to this are many. The literature has grown so that no one can seriously hope to keep up with it. The controversy accompanying its early development has largely died down and many earlier skeptics are grudgingly admitting that some good can come of these methods after all and some are, in fact, using them. Many studies in a great variety of groups of organisms are carried out using numerical methods as the principal or at least an ancillary method. And finally, mathematicians have begun to look seriously at numerical taxonomy as a worthwhile field for their investigations, with the prospect that considerable improvement and validation of these methods is in the offing.

The need for communication for those of us who work in this field is therefore very great and this is reflected in the continuing number of conferences and meetings. Last year there was an International Conference on Numerical Taxonomy in the United States. This is now matched by a similar conference under the auspices of the University of St. Andrews.

An important aspect of the development of the field has been its wide applicability to numerous scientific and scholarly disciplines. It is therefore of especially great importance that there be conferences such as these so that persons working in diverse fields are able to exchange ideas and experiences which often are, unbeknown to them, quite parallel.

Regrettably, personal factors and the tight American science budget for this year makes it impossible for me to attend this meeting. However, I look forward with great interest to the publication of its findings and wish the organisers and participants much success in their endeavors.

Robert R. Sokal

MAPPING DIVERSITY: A COMPARATIVE STUDY OF SOME NUMERICAL METHODS

A. J. Boyce

Department of Biological Sciences, University of Surrey

Introduction

The taxonomist who wishes to use numerical methods in the study of similarities and differences among organisms and for the construction of classifications now has available to him a very great variety of methods of measuring similarity and of analysing matrices of similarity values. In order to choose between these methods he needs to know their relative merits and the kinds of taxonomic information they produce. The purpose of this contribution is to consider some of the numerical methods which have been suggested for use in taxonomy and to compare the various methods in action, with an examination of the differences in outcome and a consideration of possible reasons for these differences.

The various stages in the application of numerical methods are summarised in Fig. 1. In this particular contribution, three of these stages are considered - the measurement of similarity and the analysis of matrices of similarity values by means of cluster analysis and principal components analysis.

Measurement of similarity

A geometrical model is helpful in the interpretation of coefficients of similarity. The forms studied are thought of as points lying in a multidimensional space, the axes of which correspond to the characters on which the comparisons are based. The relative positions of the points in this "character space" are determined by the particular character values possessed by each form (see Fig. 2).

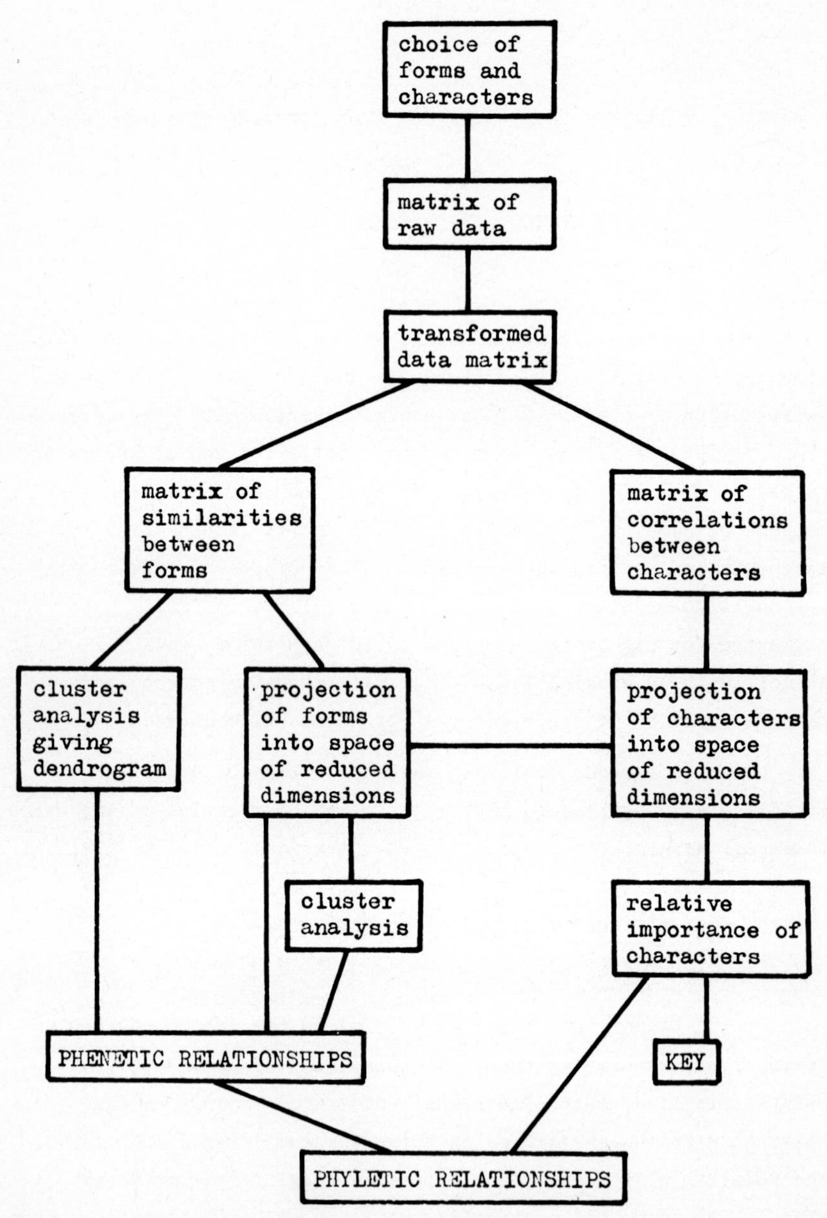

Fig. 1. Inter-connections of the stages in the numerical analysis of taxonomic relationships

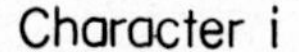

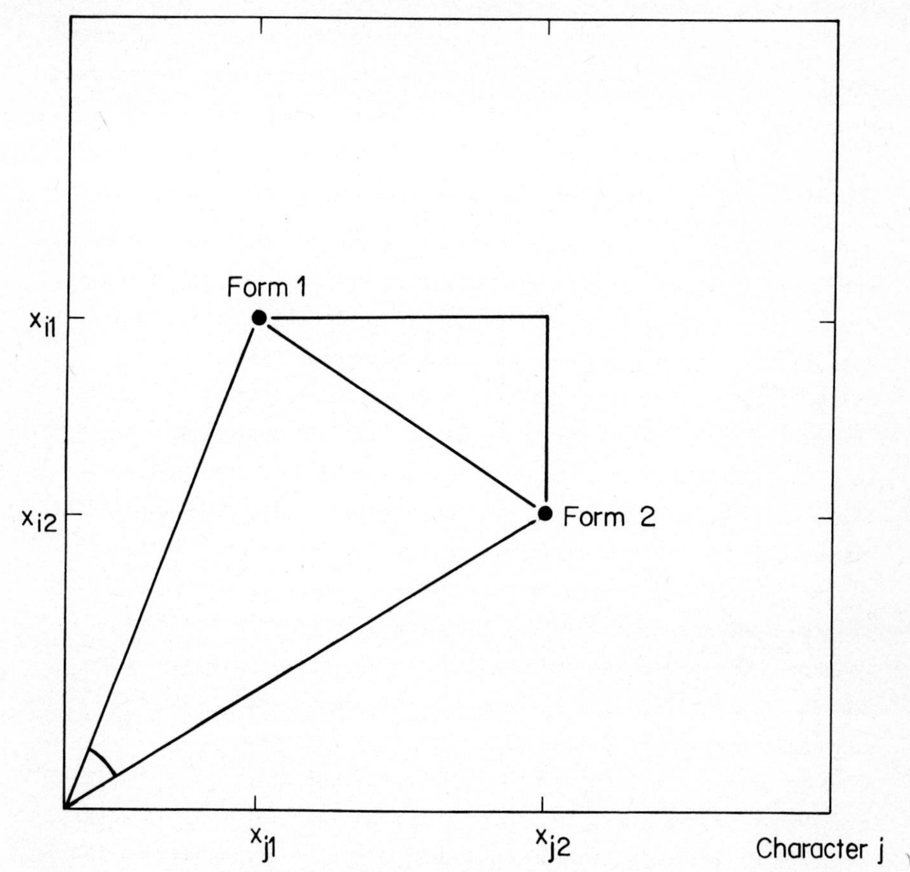

Fig. 2. Diagram to illustrate the geometrical interpretation of similarity between two forms. The upper line joining the two forms is the "city block" distance between them

The coefficients considered here are of two kinds. The first consists of coefficients which measure the angular separation of the lines connecting a pair of points to the origin of the character space; the second consists of those coefficients which measure the distance (suitably defined) between a pair of points.

Angular coefficients

The use of the angle as a measure of resemblance was first suggested by Bhattacharyya (1946). More recently, Edwards and Cavalli-Sforza (1964), in a study of relationships among human populations, measured resemblance by the angles between points (representing the populations) on the surface of a unit hypersphere. These workers measured the angle by its cosine and then transformed the cosine - Bhattacharyya to the angle itself and Edwards and Cavalli-Sforza to a distance ($d^2 = 2 - 2\cos\theta$). Here, the use of the cosine as a measure of resemblance is considered.

When the characters used in making a comparison are mainly morphological, the taxonomist frequently wishes to separate the resemblance between forms into size and shape components. One measure of size is suggested by the geometrical model. The distance of a point from the origin of the character space may be thought of as a measure of the size of the form represented by that point. Two points at different distances along the same line through the origin represent two forms which differ only in size and, conversely, two points which are not on the same line but at equal distances from the origin represent two forms which differ only in shape. Two forms will lie on the same line only if the character values of one are equal to those of the other multiplied by some constant factor (*i.e.* if $X_{i1} = k.X_{i2}$ for all i, where X_{i1}, X_{i2} represent the values of the i-th character for forms *1* and *2*). This kind of size difference is thus a *proportional* difference and it is clear that if two forms do lie on the same line through the origin, and thus differ in size only, the cosine of the angle between the lines joining them to the origin will be one, indicating complete similarity. The cosine can therefore be interpreted as a measure of similarity which ignores proportional size differences.

The term "size difference" may also be used to describe the situation in which the character values of one form differ from those of another by some constant amount (*i.e.* if $X_{i1} = X_{i2} + b$, for all i). This is an *additive* size difference. The second angular coefficient considered here, the product moment correlation coefficient, is one which ignores both additive and proportional size differences. It is equal to the cosine of the angle between two forms when the character values of the first form are expressed as deviates from the mean of all the character values of that form and the character values of the second form are expressed as deviates from the mean of its character values. The correlation coefficient has been suggested as a measure of resemblance simply on the grounds that it is a measure of covariation - if the changes in value from character to character in one form are paralleled by those in another, the coefficient will have a high, positive value, indicating a high degree of similarity. However, as Minkoff (1965) points out, the taxonomic use of the correlation coefficient as a measure of resemblance does require that all characters have the same directional and dimensional properties. This condition is fulfilled when studies of resemblance are based on morphology and the forms are characterised by linear dimensions. It is in such studies that problems of size and shape resemblance arise and in which the properties of the correlation coefficient with respect to size are most welcome.

Distance coefficients

Several distance coefficients have been proposed as measures of similarity. One of the simplest of these is the Mean Character Difference (MCD) whose use is discussed by Cain and Harrison (1958) and Huizinga (1962, 1965). Related to this is the "Taxonomic" or Mean Square Distance (MSD) and its square root, the modern use of which as measures of similarity is discussed by Penrose (1954) and Sokal (1961). These coefficients are related to a class of distance functions whose general formula is:

$$d_r(j,k) = \left(\sum_{i=1}^{n} |X_{ij} - X_{ik}|^r \right)^{\frac{1}{r}}$$

where X_{ij} is the value of character i for form j. In taxonomic use, the distance is usually multiplied by $n^{-1/r}$ to take into account the fact that the number of characters may vary from comparison to comparison. d_1 is the so-called "city-block" distance (see Fig. 2) and d_1/n is the Mean Character Difference. d_2 is the ordinary Euclidean distance and $d_2/n^{\frac{1}{2}}$ is the square root of the Mean Square Distance.

The MCD and MSD are measures of overall resemblance. The MSD, however, can be partitioned into size and shape components.

Penrose (1954) defines the size of a form as the mean of its character values and the size distance between two forms as the square of the difference in their sizes. When this distance is subtracted from the Mean Square Distance, a term remains which is the variance of the differences in character values of the two forms being compared. Penrose suggests that this variance (which he calls the Shape Distance) be used as a measure of the difference in shape between the two forms on the grounds that it ignores size differences. However, as Rohlf and Sokal (1965) point out, the Shape Distance only ignores additive size differences.

The five coefficients considered here thus react differently to the size component of resemblance and their properties are summarised in the following table:

Coefficient	Ignores: Proportional differences in size	Ignores: Additive differences in size
Mean Character Difference	No	No
Mean Square Distance	No	No
Shape Distance	No	Yes
Cosine	Yes	No
Correlation Coefficient	Yes	Yes

The formulae used in calculating these measures of similarity are given in Table I.

The search for group structure

The information about the pattern of relationships among the forms is stored in the similarity matrix. Since the numbers of forms and characters are usually large, this pattern is multi-dimensional and there is a need, therefore, for methods of analysing matrices of similarity values in order to represent the pattern of relationships in a more easily understood form. The two main types of method, cluster analysis and principal components analysis, are both concerned with reducing the number of dimensions needed to represent relationships.

Notation: X_{ij} is the value of character i for form j

n is the number of characters used in a particular comparison

Cosine of angle:

$$\cos\theta_{jk} = \frac{\sum_{i=1}^{n} X_{ij} X_{ik}}{\left(\sum_{i=1}^{n} X_{ij}^2\right)^{\frac{1}{2}} \left(\sum_{i=1}^{n} X_{ik}^2\right)^{\frac{1}{2}}}$$

Correlation coefficient:

$$r_{jk} = \frac{\sum_{i=1}^{n} X_{ij}X_{ik} - \frac{1}{n}\left(\sum_{i=1}^{n} X_{ij}\right)\left(\sum_{i=1}^{n} X_{ik}\right)}{\left(\sum_{i=1}^{n} X_{ij}^2 - \frac{1}{n}\left(\sum_{i=1}^{n} X_{ij}\right)^2\right)^{\frac{1}{2}} \left(\sum_{i=1}^{n} X_{ik}^2 - \frac{1}{n}\left(\sum_{i=1}^{n} X_{ik}\right)^2\right)^{\frac{1}{2}}}$$

Mean square distance:

$$d_{jk}^2 = \frac{1}{n}\left(\sum_{i=1}^{n} X_{ij}^2 + \sum_{i=1}^{n} X_{ik}^2 - 2\sum_{i=1}^{n} X_{ij}X_{ik}\right)$$

Mean character difference:

$$MCD_{jk} = \frac{1}{n}\sum_{i=1}^{n} \left|X_{ij} - X_{ik}\right|$$

Size difference:

$$Q_{jk}^2 = \frac{1}{n^2}\left(\sum_{i=1}^{n} X_{ij} - \sum_{i=1}^{n} X_{ik}\right)^2$$

Shape difference:

$$Z_{jk}^2 = \frac{n}{n-1}\left(d_{jk}^2 - Q_{jk}^2\right)$$

Table 1. Formulae used in calculating measures of similarity.

Cluster analysis

The cluster methods considered here are agglomerative methods, *i.e.* those in which each sub-group in an hierarchical classification is built up from the union of two smaller groups. Many different agglomerative clustering methods have been proposed but amongst the most widely used are the "group" methods proposed by Sokal and Michener (1958). The simplest of these are the pair-group methods, in which the new groups produced at any stage in the clustering process contain only two members. These groups are produced by pairing those individual forms, or groups of forms produced at an earlier stage in the clustering process, for which the relation "is most similar to" is reflexive.

The main problem which arises in the use of pair-group methods is that of choosing a suitable definition of the similarity between groups. Two solutions to this problem are considered here. One, the average pair-group method discussed by Sokal and Sneath (1963), measures the similarity between two groups as the arithmetic mean of the similarities between the individuals which make up the two groups. In one version of this method, the similarity between two groups is a weighted average of the similarities between the members of the group. That is, S_{ab}, the similarity between groups a and b is given by:

$$S_{ab} = \sum_i \sum_j w_i w_j S_{ij} / \sum_i \sum_j w_i w_j$$

where S_{ij} is the similarity between the i-th member of group a and the j-th member of group b and w_i, w_j are the weights given to these members. These weights are usually chosen to give greater weight to forms which enter groups late in the clustering process and in the method considered here are chosen as follows - unit weights are given to the two individuals which make up the original nucleus of the group; as new forms or groups of forms enter the group they are given weights equal to the sum of the weights of the forms already in the group. In the second version of the average pair-group method considered here, the unweighted version, $w_i = w_j = 1$, for all i and j.

The second solution to the problem of measuring similarity between groups uses the distance between the centroids of the groups. This distance (d_c) can be expressed (Boyce, 1965) in terms of the distances among the members of the two groups as follows:

$$d_c = \bar{B} - \frac{t_1 - 1}{2t_1}\bar{W}_1 - \frac{t_2 - 1}{2t_2}\bar{W}_2$$

where $\bar{B}$ is the mean of the t_1t_2 squared distances between the t_1 members of the first group and the t_2 members of the second group; $\bar{W}_1$ is the mean of the $\frac{1}{2}t_1(t_1-1)$ squared distances within the first group and $\bar{W}_2$ is the mean of the $\frac{1}{2}t_2(t_2-1)$ squared distances within the second group. No account is taken of the order in which forms link into groups and the centroid method is thus an unweighted method. It differs from the average method in that the distance (*i.e.* the similarity) between two groups is a function not only of the average distance between the groups but also of the average distance within each group.

The clustering techniques based on simple averages are generally applicable to both angular and distance coefficients. However, in order to apply the centroid method to angular coefficients these must first be transformed into distances. The transformation used here is:

$$d^2_{jk} = 2(1 - a_{jk})$$

where a_{jk} is either the cosine or the correlation between forms j and k.

Principal Components Analysis

Sokal (1958) appears to have been the first to employ principal components analysis in the study of tables of similarity values. Sokal applied the method only to the analysis of matrices of correlation coefficients but, as Gower (1966) has shown, it is possible to use it in the analysis of a wide variety of similarity coefficients although certain kinds, for example distance co-efficients, may need to be transformed first.

The use of principal components analysis is as follows. The latent vectors of a matrix of similarity values are extracted and the i-th components of the latent vectors are used as the co-ordinates of a point Q_i in an n-dimensional space, where n is the number of individual forms between which the similarities are calculated. The distance, in this space, between points Q_i and Q_j is then given by:

$$d_{ij}^2 = \sum_{r=1}^{n} c_{ir}^2 + \sum_{r=1}^{n} c_{jr}^2 - 2\sum_{r=1}^{n} c_{ir}\, c_{jr}$$

where c_{ir}, c_{jr} are the i-th and j-th components of the latent vector $\underline{c}_r$. If the latent vectors are normalised so that the sums of squares of their elements are equal to their corresponding latent roots then the similarity matrix, $\underline{S}$, is given by:

$$\underline{S} = \underline{c}_1\, \underline{c}_1' + \underline{c}_2\, \underline{c}_2' + \ldots + \underline{c}_n\, \underline{c}_n'$$

Thus $$S_{ii} = \sum_{r=1}^{n} c_{ir}^2 \text{ and } S_{ij} = \sum_{r=1}^{n} c_{ir}\, c_{jr} \,.$$

Therefore, when the latent vectors of the similarity matrix are normalised so that $\Sigma\, c_{ir}^2 = \ell_r$, where ℓ_r is the r-th latent root, the distances between the points Q_i and Q_j are related to the original similarity values as follows: (Gower, 1966):

$$d_{ij}^2 = S_{ii} + S_{jj} - 2S_{ij} \,.$$

The points Q_i often have the right sort of metric properties for representing the pattern of relationships between the forms. For example, in the case of angular coefficients where $S_{ii} = S_{jj} = 1$

$$d_{ij}^2 = 2(1 - S_{ij}).$$

If a principal components analysis is carried out on a matrix of distance coefficients, one or more of the latent roots may be negative and, if so, one or more of the co-ordinate axes imaginary. In this study, the transformation used to overcome this difficulty is:

$$S_{ij} = 1 - k.S'_{ij}$$

where S'_{ij} is the distance coefficient (MCD, MSD or Shape Distance) between forms i and j, and k is chosen to ensure that S_{ij} is never negative. Using this transformation, the distance between points Q_i and Q_j is given by:

$$d^2_{ij} = 2k.S_{ij} .$$

An exact representation of the relationships between forms can be obtained only by using all latent vectors in determining the co-ordinates of the points Q_i. However, when some of the latent roots are small a good representation can be obtained in a reduced number of dimensions, for, if a latent root ℓ_r is small then the contribution $(c_{ir} - c_{jr})^2$ to the distance between Q_i and Q_j will also be small. Even if ℓ_r is large but the c_{ir} corresponding to it are not very different then $(c_{ir} - c_{jr})^2$ will still be small. The only co-ordinates, therefore, which contribute much to the distances are those with large latent roots and wide variation in the elements of their vectors. In many applications it is found that the distances can be adequately expressed in terms of two or three such vectors.

The material used in making the comparisons

In order to compare the relative merits of the methods and the kinds of taxonomic information they produce, the phenetic relationships of four groups of hominoids were studied. These were the three genera of large apes (*Pan*, chimpanzees; *Pongo*, orang-utans; *Gorilla*, gorillas) and modern *Homo sapiens*. Each group was represented by a male, female and juvenile specimen. In addition, several fossil forms were included to allow comparisons to be made between the patterns of relationship of the fossil forms (with each other and with the four modern groups) revealed by the different methods. Only skulls were considered in this study since these were thought to provide sufficient information about the phenetic relationships of the forms to allow meaningful comparisons to be made between the different methods. The fossil skulls were chosen to sample each of the main evolutionary stages leading to modern man: the australopithecine stage was represented by a cast of the infant skull from Taung, a cast of an almost complete skull from Sterkfontein ("Plesianthropus") and a cast of a skull from Swartkrans ("Paranthropus"). The erectus stage was represented by a cast of the skull of Pekin man (a reconstructions by Weidenreich based on skulls from Choukoutien) and the hominine stage by casts of skulls from Broken Hill (Rhodesian man), La Chapelle aux Saints, Mount Carmel (skull 5 from the Skhul cave) and Steinheim. The twenty skulls are listed in Table 2. Each skull was given a reference number, this appears in the right-hand column of the table.

The skull was divided into fourteen functional areas and characters were taken from each area. An attempt was made to characterise those aspects within each functional area which varied over the forms considered, and 99 characters were needed to do this. Further details of the characters are not given here since the numerical methods are compared using the same set of characters and the effects of using different characters are not considered. The criteria used in choosing the characters are discussed in Boyce (1964, 1965) and lists of the characters and the raw character values are given in Boyce (1965).

Comparisons between the methods

Similarity coefficients were calculated between each pair of forms and since 20 forms were used, 190 comparisons were made. The coefficients were calculated from transformed character values. The particular transformation used in this study is standardisation in which the values of a particular character are transformed as follows:

$$X'_{ij} = (X'_{ij} - \bar{X}_i)/s_i$$

PONGIDAE	Pan ♂	1
	Pan ♀	2
	Pongo ♂	3
	Pongo ♀	4
	Gorilla ♂	5
	Gorilla ♀	6
	Pan j	7
	Pongo j	8
	Gorilla j	9
AUSTRALOPITHECUS	Taung j	10
	Sterkfontein	11
	Swartkrans	12
HOMINIDAE	Pekin (Choukoutien)	13
	La Chapelle aux Saints	14
	Rhodesian (Broken Hill)	15
	Skhul (Mount Carmel)	16
	Steinheim	17
	Homo sapiens ♂	18
	Homo sapiens ♀	19
	Homo sapiens j	20

Each skull was given a reference number and these numbers appear in the right-hand column. j = juvenile.

Table 2. List of hominoid skulls measured.

where X'_{ij} is the raw value and X'_{ij} the transformed value of character i for form j; $\bar{X}_i$, s_i are the mean and standard deviation of character i calculated over all the forms studied. This method is discussed by Rohlf and Sokal (1964) and Boyce (1965).

In the following sections, methods of analysing matrices of similarity values are compared first. Coefficients of similarity are then compared using one particular method of cluster analysis.

Cluster Analysis

The three methods compared here are the weighted pair-group method using averages (WPGMA), the unweighted pair-group method using averages (UPGMA) and the unweighted pair-group method using centroids (UPGMC). The methods are applied to matrices of correlation coefficients and mean square distances.

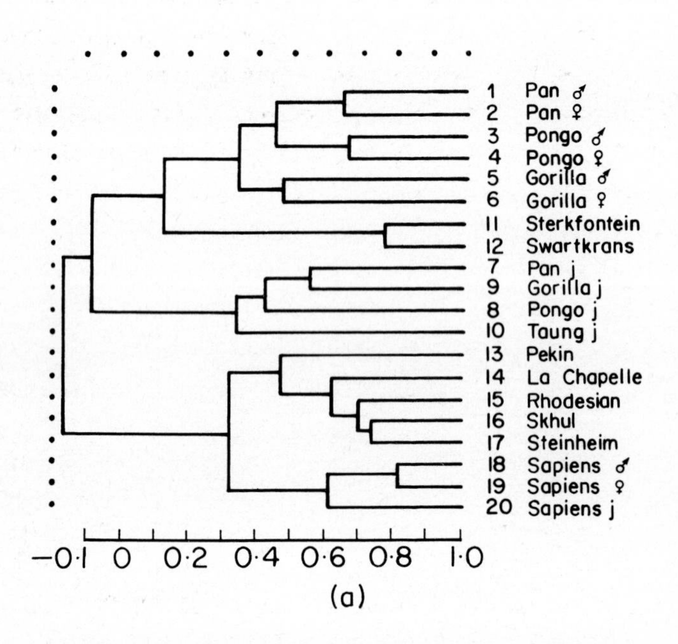

Fig. 3. Comparison between average and centroid methods of cluster analysis

(a) Dendrogram produced by weighted average pair group method

Fig. 3 shows the results of applying the three methods to a matrix of correlations between the forms. In the dendrograms the twenty forms fall into two main groups - the first consists of the adult and juvenile apes and the Australopithecines (forms 1-12) and the second consists of the fossil and modern hominids (13-20). Within the first group there is a division into three sub-groups: the adult apes (1-6), the juvenile apes (with the juvenile Australopithecine from Taung, 7-10) and the adult Australopithecines (11,12). Within the hominid group there is a division into modern (18-20) and fossil forms (13-17).

The overall patterns of relationship produced by the three methods are very similar and there are no topological differences between the dendrograms based on averages (Figs. 3a & 3b), although the levels at which corresponding stems join do differ. In the average methods, as the groups link up, the average similarity always decreases. In the centroid method, however, the distance between centroids sometimes decreases and hence the similarity between groups increases. "Reversals" thus occur on the dendrogram.

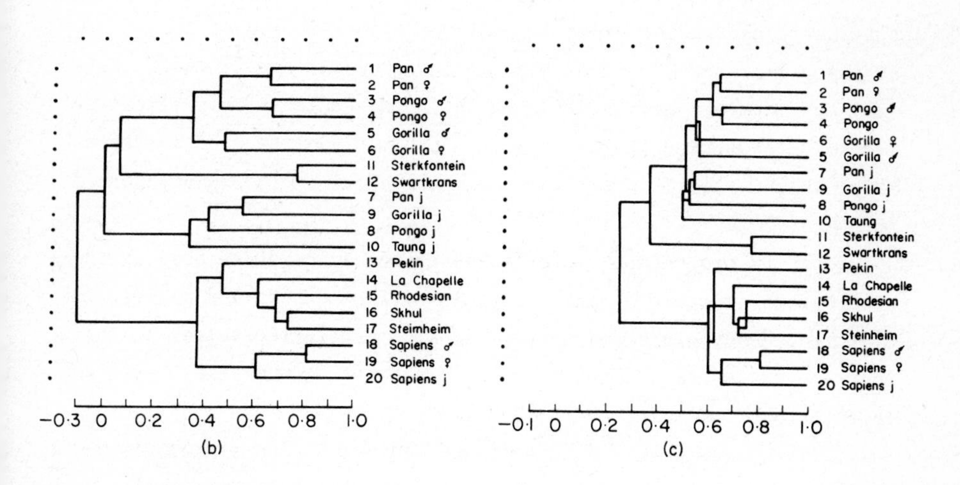

Fig. 3. (continued)

(b) Dendrogram produced by unweighted average pair group method

(c) Dendrogram produced by unweighted pair group method using distances between centroids

Dendrograms based on correlations between hominoid skulls calculated from standardised character values

When weighted and unweighted methods are applied to a matrix of Mean Square Distances, more marked differences in outcome appear. Fig. 4a is a dendrogram produced by the unweighted average method and Fig. 4b is a dendrogram produced by the weighted average method. The most striking difference in the patterns of relationship is in the relative positions of the fossil hominids and the juvenile apes. The dendrogram based on the unweighted method (Fig. 4a), the group of juvenile apes and Taung (7-10) links with the group of adult apes and adult Australopithecines (1,2,4,6, 11,12) and the group of fossil hominids (13-17) links with the modern *Homo sapiens* group (18-20). In the dendrogram based on the weighted method, however, the pattern is reversed. The reason for the difference lies in the relative weights given to the members of these groups by the two methods. If the "outlying" members of the four groups are considered, it is found that the outliers of the hominid group (13,15) face (to continue the geometric analogy) those of the ape-Australopithecine group (11,12) while the outlier of the *H. sapiens* group (20) faces that of the juvenile ape-Taung group (10) and the very much greater weight attached to these outliers in the weighted method is sufficient to produce a pattern of relationships which is quite the reverse of that shown by the unweighted method.

In order to measure the degree of concordance between the dendrograms produced by different clustering methods and between the different dendrograms and the matrices of similarity values on which they are based, an extension of the method of "cophenetic correlations" of Sokal and Rohlf (1962) is used. The *cophenetic value* of two forms in a dendrogram is defined as the similarity value at which the stems from them join. Cophenetic values between all pairs of forms are calculated and the values obtained are compared with the similarity values on which the dendrogram is based or with cophenetic values obtained from another dendrogram. The concordance is measured by the correlation between the two sets of values.

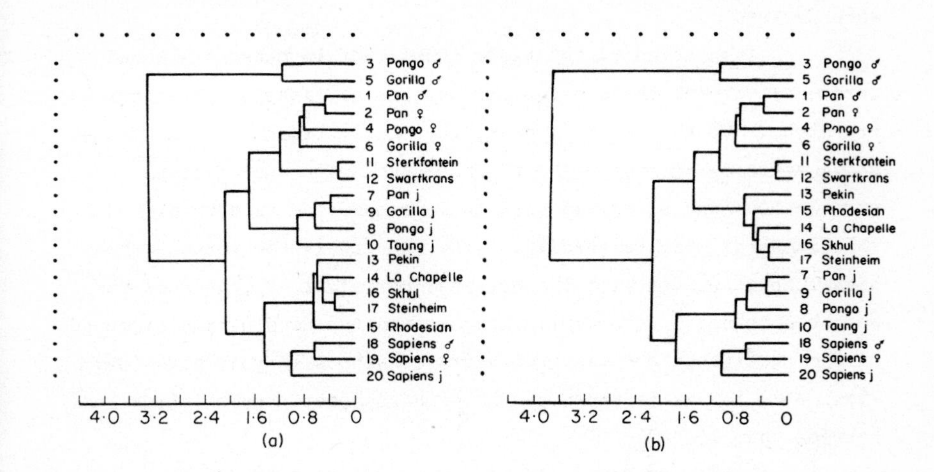

Fig. 4. Comparison between unweighted and weighted methods of cluster analysis

(a) Dendrogram produced by unweighted average pair group method

(b) Dendrogram produced by weighted average pair group method

Dendrograms based on mean square distances calculated from standardised character values

The cophenetic correlations obtained by applying this technique to the three dendrograms based on correlations between forms (Fig. 3) are shown below:

	r	WPGMA	UPGMA	UPGMC
r	-	0.861	0.890	0.869
WPGMA		-	0.982	0.898
UPGMA			-	0.959
UPGMC				-

The abbreviations are those used on p.14; r denotes the original matrix of correlations between forms from which the dendrograms were derived.

Inspection of the table shows that in general the concordances between the dendrograms and the original correlations are high. Of the three methods, the unweighted pair-group method using averages has the highest concordance with the original correlations and so may be said to reproduce the pattern of relationships most faithfully. The two unweighted methods show higher concordances with the original correlations than does the weighted method. The similarity between the dendrograms produced by the unweighted and weighted average methods is very high (see Figs. 3a & 3b) as is shown by the high cophenetic correlation between them (0.982).

Sokal and Rohlf (1962) compared four methods of cluster analysis - two variable and two pair group methods. The pair group methods were both weighted methods, one expressed similarity between groups by Spearman's sums of variables method (see Sokal and Sneath, 1963, p.183), the other by the average similarity between the groups (WPGMA). Of the four methods, they found that the pair group method using averages gave the highest correlation with the original similarity values. They found a correlation of 0.86 for this method (this happens to agree closely with the value found here). Unfortunately, since Sokal and Rohlf did not consider any unweighted pair group methods, further comparisons with their results cannot be made.

Principal Components Analysis

The second solution to the problem of analysing matrices of similarity values is the method of principal components analysis. The method is used here to analyse matrices of correlations and mean square distances between forms.

In the first analysis, latent roots and vectors were extracted from a matrix of correlations and the vectors were normalised so that the sum of squares of the elements of each vector was equal to the corresponding latent root. In Fig. 5 the first three normalised vectors are used as co-ordinate axes and the positions of points representing the forms are shown with respect to the first and second vectors in Fig. 5a and the first and third vectors in Fig. 5b. The first three vectors were also used in the construction of a model. One view of this is shown in Fig. 6. An examination of these figures shows that the forms fall into two quite distinct clusters. The apes and Australopithecines (forms 1-12) lie in the upper half of the space and are quite separate from the fossil and modern hominids (13-20) which lie in the lower half. Within the ape-Australopithecine cluster there is a division into three sub-clusters (as in the dendrograms based on correlations, see Fig. 3). These are the juvenile forms, the adult apes and the adult Australopithecines. Within the hominid cluster there is a division into fossil and modern forms.

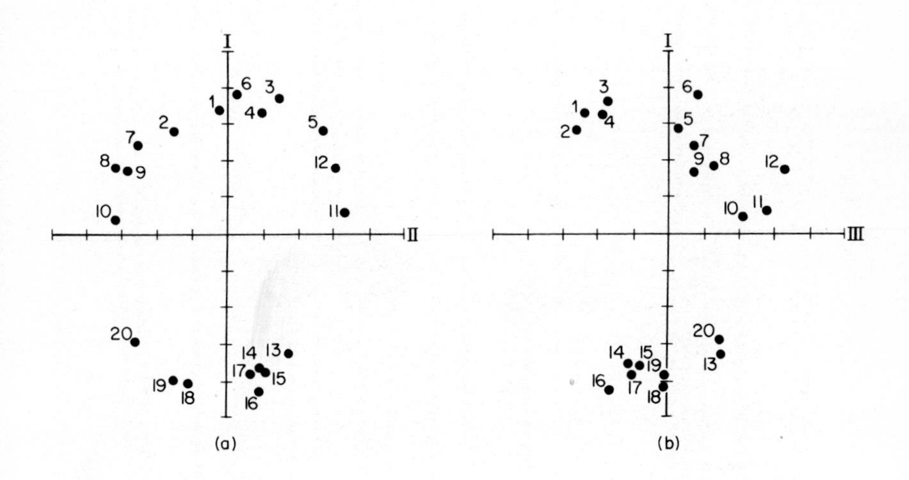

Fig. 5. Diagram to show the results of a principal components analysis of correlations between hominoid skulls based on standardised character values.

(a) Forms plotted with respect to the 1st and 2nd latent vectors.

(b) Forms plotted with respect to the 1st and 3rd latent vectors.

Reference numbers as in Table 2.

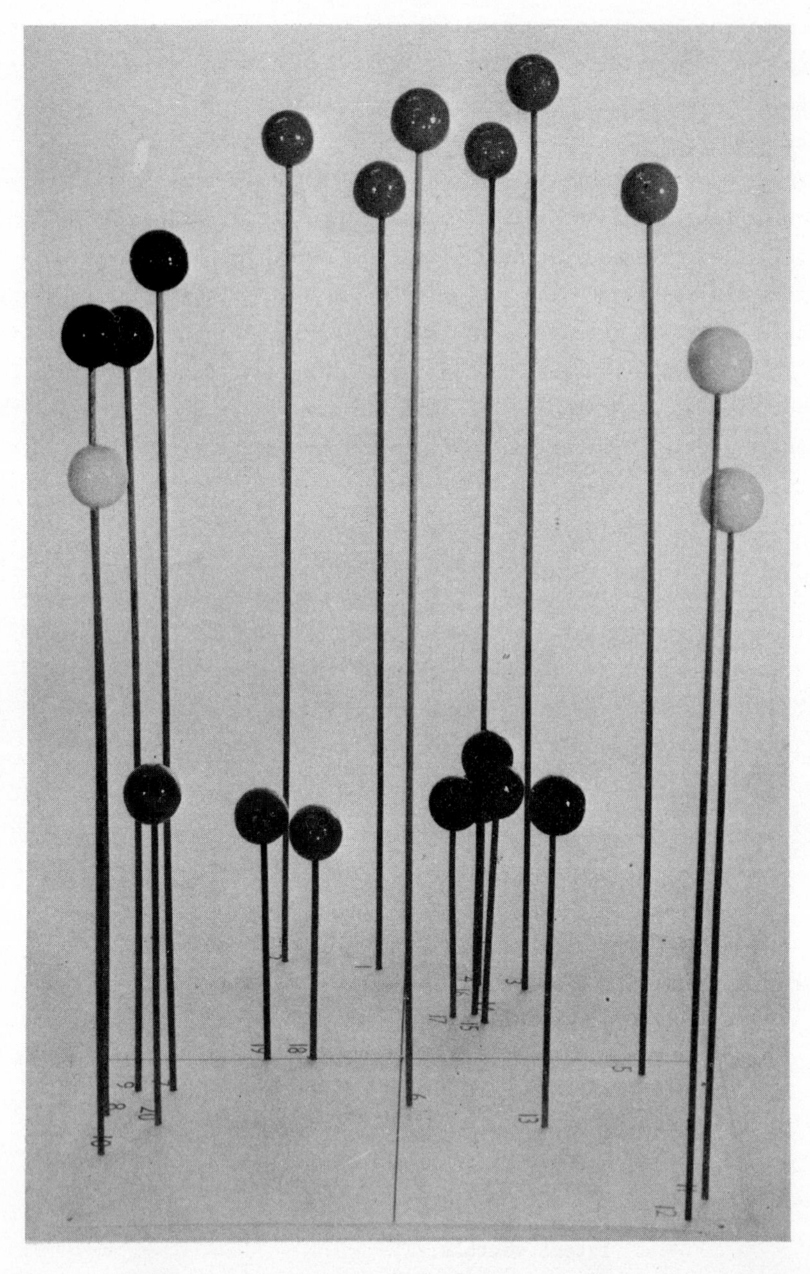

Fig. 6. Model to show the results of the principal components analysis of hominoid correlations. The vertical axis corresponds to the 1st latent vector and the horizontal axes to the 2nd and 3rd latent vectors. The model is viewed at right angles to the 2nd vector. Reference numbers on base as in Table 2.

The success of any group of vectors in reproducing the original matrix of correlations can be measured by the relative magnitude of their corresponding latent roots. In this example, approximately two-thirds (66.8%) of the total variation among the forms is accounted for by the first three latent vectors. A second method of measuring the distortion produced by the reduction of dimensions is the method of cophenetic correlations. Cophenetic values are obtained by calculating the distances between all pairs of forms in the space defined by the first three vectors. On p.17 it was shown that the distances between forms (using all latent vectors as co-ordinate axes) and the correlations between forms are related by the formulae:

$$d_{jk}^2 = 2(1-r_{jk}) \quad or \quad r_{jk} = 1 - \frac{1}{2}d_{jk}^2 .$$

If the distances based on three vectors are transformed into "correlations" by the second of these formulae, the new values can be compared with the original correlations and with cophenetic values from the dendrograms and the goodness of fit measured by the method of cophenetic correlations. The results of applying this technique are shown below:

	r	WPGMA	UPGMA	UPGMC
PCA_3	0.969	0.817	0.859	0.852

The first three vectors give a very good reproduction of the original correlations, and reproduce these more faithfully than does any of the dendrograms. Of the three dendrograms, that produced by the unweighted average method shows the best fit with the results of the principal components analysis.

Latent roots and vectors were also extracted from a matrix of mean square distances, using the transformation: $S_{jk} = 1\text{-}0.1\ MSD_{jk}$. The most marked difference between the latent vectors of the mean square distance matrix and those of the correlation matrix previously considered is that the elements of the first latent vector of the mean square distance matrix are nearly constant whereas those of the first vector of the correlation matrix show a wide range of positive and negative values. Since this vector contributes very little to the distances between points representing the forms in the space defined by the latent vectors, it was ignored and the pattern of relationships with respect to the second, third and fourth vectors was examined.

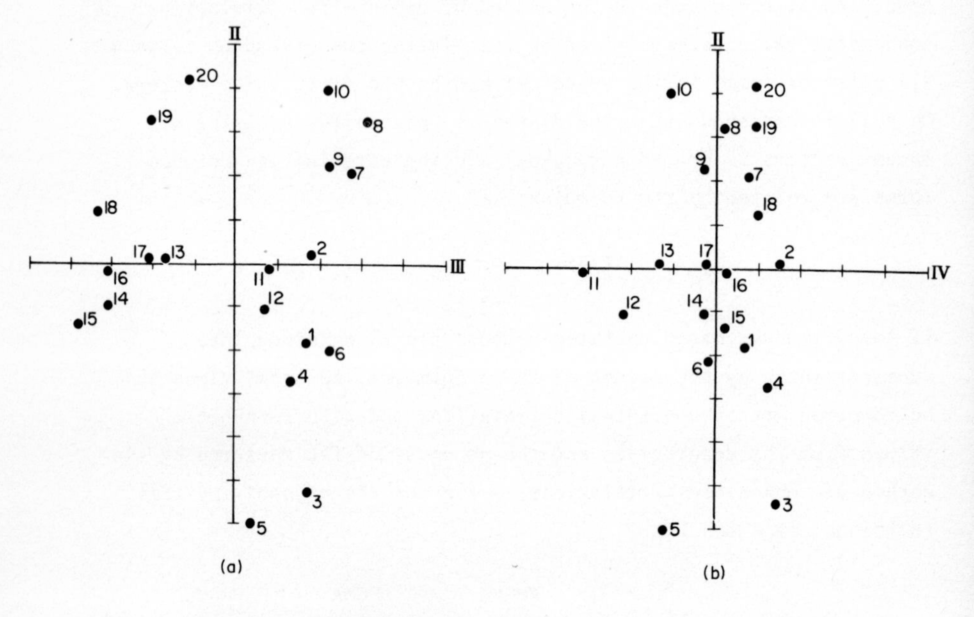

Fig. 7. Diagram to show the results of a principal components analysis of mean square distances between hominoid skulls based on standardised character values.

(a) Forms plotted with respect to the 2nd and 3rd latent vectors.

(b) Forms plotted with respect to the 2nd and 4th vectors. Reference numbers as in Table 2.

The results of this analysis of the matrix of mean square distances are shown in Figs. 7 and 8. In Fig. 7a the forms are plotted with respect to the second and third vectors; in Fig. 7b with respect to the second and fourth vectors. In Fig. 8, the plot against the second and third vectors is shown, for comparison, by the dendrogram produced by the unweighted average cluster analysis

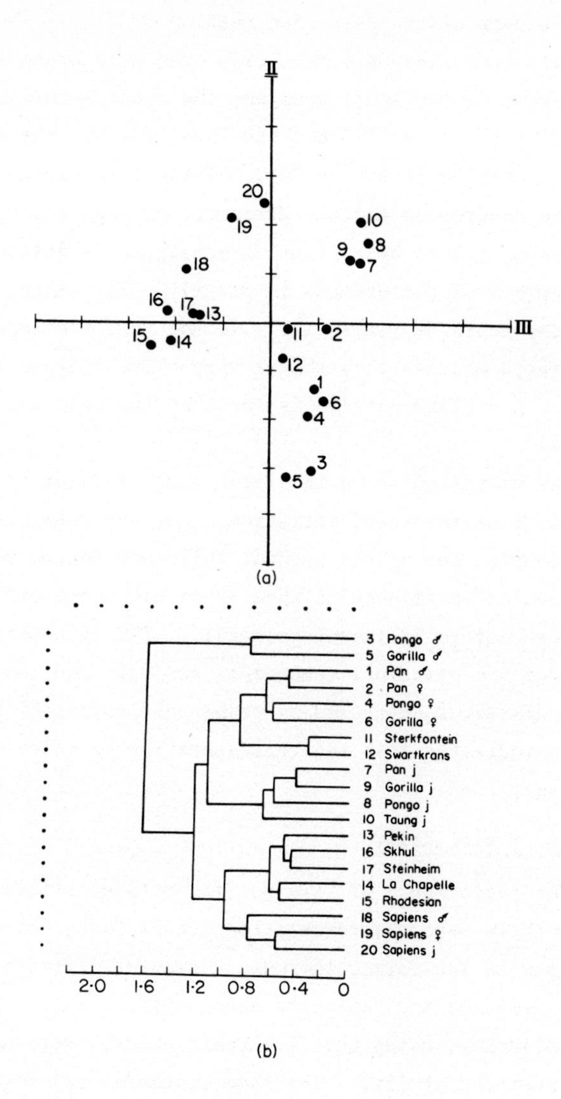

Fig. 8. Comparison between principal components analysis and cluster analysis of mean square distances

(a) Forms plotted with respect to the 2nd and 3rd latent vectors

(b) Dendrogram produced by unweighted average pair group cluster analysis.

of the mean square distances. Examination of Fig. 7 shows that the forms fall into five main clusters: the male orang and gorilla (3,5): the rest of the adult apes and the adult Australopithecines (1,2,4,6,11,12) the juvenile apes and Taung (7-10), the fossil (13-17) and modern hominids (18-20). These five clusters also appear in the dendrogram and the agreement between the results of the two methods, judged by a visual comparison, is quite good but there is a number of differences in detail. For example, in the principal components diagrams, the male and female chimpanzees are widely separated whereas they show a very close linkage in the dendrogram. A similar pattern is shown by the male and female *Homo sapiens*.

The variation among the forms accounted for by the first vector is 81.7% of the total variation. Of the remaining variation, 77.1% is accounted for by the second, third and fourth vectors. The distortion of relationships when these are shown with respect to only three vectors is therefore small. The agreement between the results of the principal components analysis and the original mean square distances, measured by cophenetic correlation, is 0.992, which indicates that the representation by these three vectors is very good.

Measurements of similarity

The performance of five coefficients of similarity was examined. These were the correlation coefficient, the cosine of the angle between two forms, the mean character difference, the mean square distance and Penrose's shape coefficient. Measurements of similarity, using the five coefficients, were made between the twenty hominoid skulls. The same character values were used in the calculation of each of the five sets of coefficients and were obtained by standardising the raw character values. The five sets of similarity values were analysed by the unweighted average pair group method of cluster analysis and the results are shown in Figs. 9 and 10.

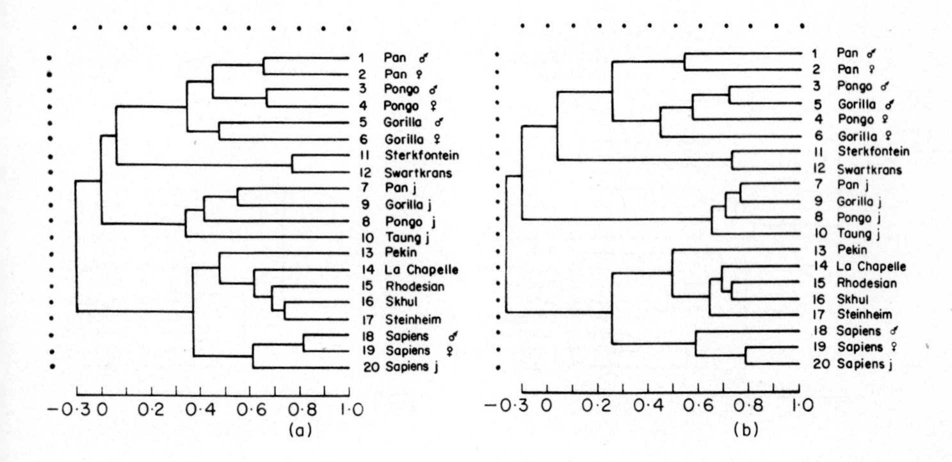

Fig. 9. Patterns of relationship shown by a) correlation coefficients and b) cosines

Dendrograms produced by unweighted average pair group cluster analysis of similarities based on standardised character values

Fig. 9 shows the pattern of relationships produced by the two angular coefficients - Fig. 9a is based on correlations and Fig. 9b on cosines. The main clusters into which the forms fall and the linkages between clusters are the same in both dendrograms. The patterns within the clusters, however, are not the same. When similarity is measured by the correlation coefficient (Fig. 9a), the male apes and the male *Homo sapiens* are most similar to their corresponding females. When similarity is measured by means of the cosine, however, the male orang and gorilla are more similar to each other and the female *H. sapiens* is more similar to the juvenile than to the male *H. sapiens*. The pattern within the fossil hominid group also varies with the coefficient used.

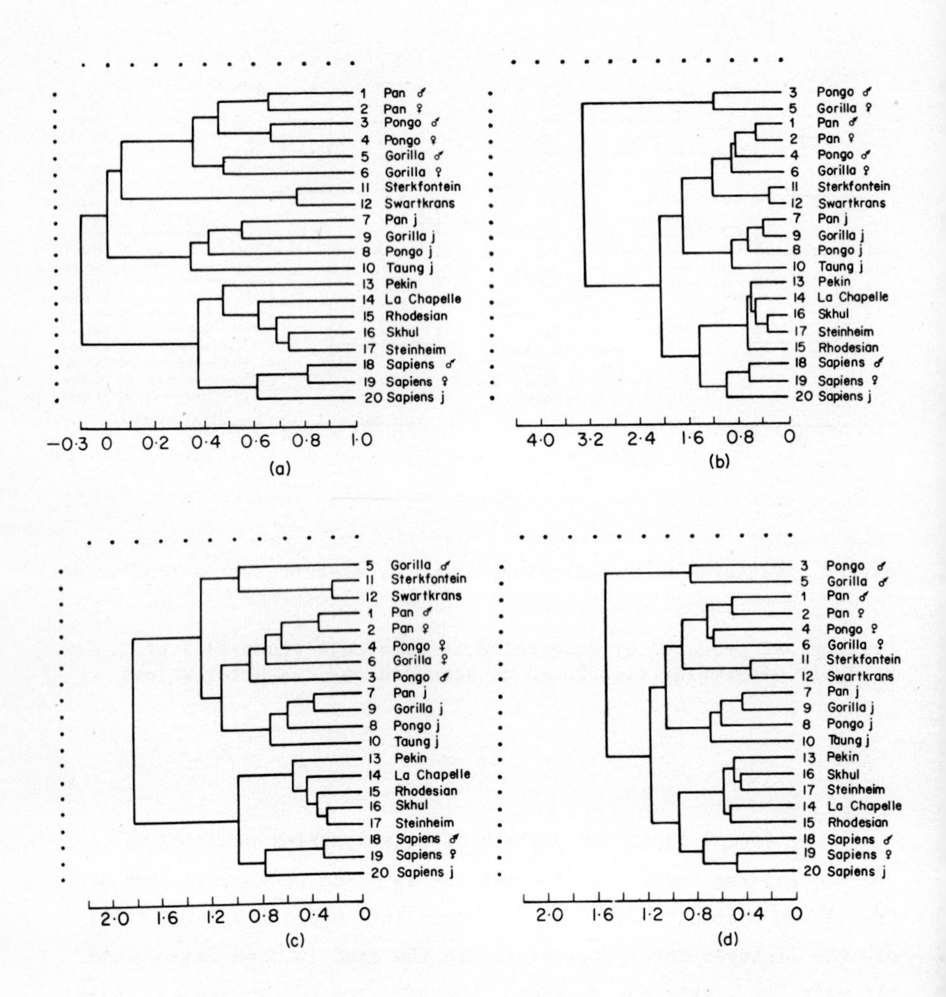

Fig. 10. Patterns of relationship shown by
a) correlation coefficients, b) mean square distances, c) Penrose's shape coefficients and d) mean character differences

Dendrograms produced by unweighted average pair group cluster analysis of similarities based on standardised character values

Fig. 10 compares the pattern shown by the correlation coefficient with those shown by the three distance coefficients and examination of the figure shows that irrespective of the coefficient used to measure similarity, the same overall pattern of relationships emerges. The apes are clearly separated from the hominids; within the pongids there is a division into juvenile and adult forms and, within the hominids, a division into fossil and modern forms. The Australopithecines always show a greater similarity with the apes than with the hominids. When similarity is measured by means of distance coefficients, however, the male orang and gorilla show less similarity with their corresponding females and in the dendrograms based on mean squared distances and mean character differences, the male orang and gorilla form a cluster which is quite separate from the other clusters.

Discussion

It is important to remember, when considering the differences between the dendrograms based on different similarity coefficients, that the pattern of relationships shown by a dendrogram is only an approximate representation of the matrix of similarity values on which it is based and that small variations in the similarity values may be sufficient to alter linkages. In assessing the relative merits of the five coefficients, therefore, great emphasis cannot be placed on the minor differences between the dendrograms. The position of the male orang and gorilla, however, shows a very striking variation among the five dendrograms and an examination of the differences between the male orang and gorilla and their corresponding females throws light on the kind of resemblance measured by the five coefficients. Orangs and gorillas show a high degree of sexual dimorphism which takes the form in the males of an overall increase in size and in exaggerated development of the sagittal and occipital crests as a result of the increase in the size of the jaws. An examination of the individual differences in the (standardised) character values of the male and female gorilla skulls confirms the conclusion that the differences between the males and females are mainly size differences. The twenty characters which show the largest absolute difference in value were examined. Of these twenty characters, which account for 43% of the total difference in character values between the two skulls, only two have a smaller value in the male. However, the differences in character values are not constant and, furthermore, are not evenly distributed over the skull but are mainly features of the jaws and of the temporal and occipital regions. The difference in size between the skulls of the male and female gorilla is thus a combination of proportional and additive size differences.

If the five coefficients are ordered according to the degree of isolation shown by the male orang and gorilla, the order is: correlation coefficient (least isolation), cosine of angle, shape coefficient, mean character difference, mean square distance. Since the difference between the male and female in the orang and gorilla is mainly a difference in size, it is reasonable to conclude that the differences between the dendrograms mainly reflect differences in sensitivity to the size component of resemblance on the part of the five coefficients. Of the five coefficients, the correlation coefficient reacts to the components of size and shape in a way most similar to that which the taxonomist usually adopts since it links together the male and female of each of the four modern groups.

The comparisons between the various methods of analysing matrices of similarity values show that there is relatively little difference in outcome between the three methods of cluster analysis. Of the three, the unweighted method based on averages (UPGMA) reproduces the relationships between the forms most accurately. Although cluster analysis produces a diagram in which the groupings are shown clearly and forcefully, the dendrogram is a poor guide to the relationships between the members of one cluster and those of another. The diagrams produced by principal components analysis reproduce the original similarity values more faithfully and give more information about the relationships of individual forms than do the dendrograms. Principal components analysis can be applied with equal success to matrices of angular or distance coefficients.

Acknowledgements

This paper is based on research carried out during the tenure of a Junior Research Fellowship at St. John's College, Oxford.

References

Bhattacharyya, A. 1946. On a measure of divergence of two multinomial populations. *Sankhya, 7* : 401-406.

Boyce, A.J. 1964. The value of some methods of numerical taxonomy with reference to hominoid classification. In: V.H. Heywood and J. McNeill (eds.), *Phenetic and phylogenetic classification* pp.47-65. Systematics Association Publication No. 6, London. 164 pp.

Boyce, A.J. 1965. The methods of quantitative taxonomy with special reference to functional analysis. D. Phil. Thesis, Oxford, vi + 190 pp.

Cain, A.J. and Harrison, G.A. 1958. An analysis of the taxonomist's judgment of affinity. *Proc. Zool. Soc. Lond.*, *131* : 85-98.

Edwards, A.W.F. and Cavalli-Sforza, L.L. 1964. Reconstruction of evolutionary trees. In: V.H. Heywood and J. McNeill (eds.) *Phenetic and phylogenetic classification* pp. 67-76. Systematics Association Publication No.6, London. 164 pp.

Gower, J.C. 1966. Some distance properties of latent root and vector methods used in multivariate analysis. *Biometrika, 53* : 325-338.

Huizinga, J. 1962. From DD to D^2 and back. The quantitative expression of resemblance. *Proc. K. ned. Akad. Wet.*, Series C, *65* : 380-391.

Huizinga, J. 1965. Some more remarks on the quantitative expression of resemblance (distance coefficients). *Proc. K. ned. Akad. Wet.*, Series C, *68* : 69-80.

Minkoff, E.C. 1965. The effects on classification of slight alterations in numerical technique. *Syst. Zool.*, *14* : 196-213.

Penrose, L.S. 1954. Distance, size and shape. *Ann. Eugenics*, *18* : 337-343.

Rohlf, F.J. and Sokal, R.R. 1965. Coefficients of correlation and distance in numerical taxonomy. *Univ. Kansas Sci. Bull.*, *45* : 3-27.

Sokal, R.R. 1958. Quantification of systematic relationships and of phylogenetic trends. *Proc. X. Intern. Cong. Entomol.*, *1* : 409-415.

Sokal, R.R. 1961. Distance as a measure of taxonomic similarity. *Syst. Zool.*, *10* : 70-79.

Sokal, R.R. and Michener, C.D. 1958. A statistical method for evaluating systematic relationships. *Univ. Kansas Sci. Bull.*, *38* : 1409-1438.

Sokal, R.R. and Rohlf, F.J. 1962. The comparison of dendrograms by objective methods. *Taxon*, *11* : 33-40.

Sokal, R.R. and Sneath, P.H.A. 1963. *Principles of numerical taxonomy*. W.H. Freeman and Company, San Francisco, 359 pp.

Discussion

Q. Wishart said that angular coefficients were strongly dependent on the choice of origin. For example, in the case of percentages, the origin could be 0%, 50% or 100%, with very different results. He asked whether Boyce had tried comparisons to test this.

A. Boyce agreed in general but pointed out that the comparisons reported in his paper were all based on standardised linear dimensions with character means as origins. He had made comparisons using unstandardised characters but found these much harder to interpret.

Q. Hope thought that numerical taxonomists had tended to ignore the earlier work by psychologists, who were already familiar with the controversy over the relative merits of analyses of correlations between objects versus correlations between characters. How could one use a "mean" of disparate characters of one object? And if not the mean, how much less the correlations.

A. Boyce felt the relative merits of the two techniques still needed further investigation to establish their usefulness in numerical taxonomy. He thought that his study of the correlations between objects (the skulls) was legitimate since the characters used to describe the objects were all standardised and most were linear dimensions, measured in the same units.

Q. Greenwood asked whether, if ones aim was to elucidate phylogeny, one ought not to look to species as ones objects, rather than individuals.

A. Boyce said there were many questions for which individuals were more relevant, particularly in palaeoanthropology, where species boundaries were very controversial and often based on small numbers of individuals. A quantitative, more precise mapping of similarities and differences could help to resolve these controversies.

Q. Bisby asked how the characters were chosen.

A. Boyce said they were selected so as best to differentiate the skulls.

Q. Bisby asked if they were mainly traditional.

A. Boyce said they were not, since traditional ones had no functional significance and could not easily be applied to apes. The characters were newly defined for the investigation.

THE USE OF GRAPHICAL METHODS IN CLASSIFICATION

R. M. M. Crawford

Department of Botany, The University, St. Andrews

In ecology use has always been made of maps and diagrams to express the end results of vegetation surveys. It is, therefore, not surprising that the development of ordination techniques (the representation in 2 or 3 dimensions of the floristic affinities between samples of vegetation) has also been exploited by this branch of biology. Maps and ordination diagrams are graphical attempts to put on paper the changes that take place in vegetation in response to an environmental gradient. In making a vegetation map it is physical gradients such as altitude, aspect and geology that are related to the plant cover. In the ordination diagram the gradient that will be extracted will not necessarily be connected with the spatial dimensions of the habitat, but may be some factor, such as soil acidity or moisture retention which varies independently of geographical location. In the first process, map making, a classification is made and then plotted on paper. In the second process, there is no preliminary classification, but the changes in the vegetation are allowed to express themselves as a continuum.

It is unfortunate that these procedures have been regarded as totally opposed to one another, and that consequently much time has been spent on arguing their relative merits and defects. In practice the value of any classification is limited unless we have some representation of the affinities between the final groups, and for this some ordination process is necessary. Similarly, with ordination a complete continuum is rarely found and in fact the method seeks to define the existence of various clusters or classes for which some means of drawing boundaries is desirable. Ideally, for botanical data at least some compromise is needed between these two processes. This raises a number of difficulties. In both classification and ordination correlations have to be tested between, either every pair of species, or every pair of objects. There is, therefore, a quadratic relationship between the size of the survey and computing time. This is a serious obstacle in botanical surveys where both the number of objects and attributes tend to be large. This paper suggests an approach to classification that avoids the quadratic relationship between the number of species and samples and computing time and uses the information obtained from the classification to hasten the process of ordination. An attempt is also made to suggest a method of mapping vegetation which yields the maximum amount of information on vegetation and environment gradients.

To achieve this end it has been necessary to adopt an approach that differs widely from most current methods. The species of plants occurring in any sample of the vegetation are considered as integrators or symptoms of an ecological situation. In medical science groups of symptoms are classified into syndromes which are diagnostic of certain pathological conditions, so therefore can plant species be considered as symptoms of particular ecological situations. The greater the number of symptoms, the more certain is the diagnosis. Further, common symptoms are considered more informative than rare ones. Thus in botanical terms, it is the species which occur frequently together which determine an ecological group and not those that are frequent but isolated or infrequent yet occurring in floristically rich areas.

Thus to assess the potential of a species for forming a group we use the product statistic,

$$W_x^1 = P_x V_x$$

which is obtained from

$$P_x = \frac{\text{The number of occurrences of species } X}{\text{The number of quadrats in the population sample}}$$

and

$$V_x = \frac{\text{The total number of species occurrences in those quadrats containing } X}{\text{The frequency of } X}$$

The value of W_x^1 so obtained is termed the group element potential (GEP) value of species. It follows that if this value is known for each species we may use it for constituting a classification of the quadrat attributes. For this the sum of all the GEP values of the species occurring the quadrat are taken, and represented as S_j the set element potential (SEP). It is now possible to test the effect of dividing the set of samples on the presence and absence of each species in turn on the set element potential (SEP). (for details see Crawford and Wishart, 1967). Thus if there are N species in the survey only N tests need be made for each division of the data. The usual quadratic relationship between species number and computing time has therefore been avoided. When the subsets have reached an arbitrarily selected level of species coincidence - ϕ -(see Crawford and Wishart, 1967) the division is terminated. The results can be represented hierarchially as in fig. 1.

The resulting groups from this classification may be mapped in the conventional manner. However, if the vegetation has been sampled systematically on a grid it is possible to plot by a contouring process the potential of each quadrat for belonging to any of the final sets. This process avoids the error of assuming that discrete boundaries exist between one vegetation type and another. Frequently a view of vegetation in the field suggests the existence of abrupt boundaries. When this is examined by such a contouring process (Crawford and Wishart 1968) as this it can be seen (fig.2) that the eye has been misled by the distribution of the physiognomically dominant species. This retrieval of the quadrat data after classification has allowed a more accurate representation to be made of the changes in the vegetation in response to an environmental gradient.

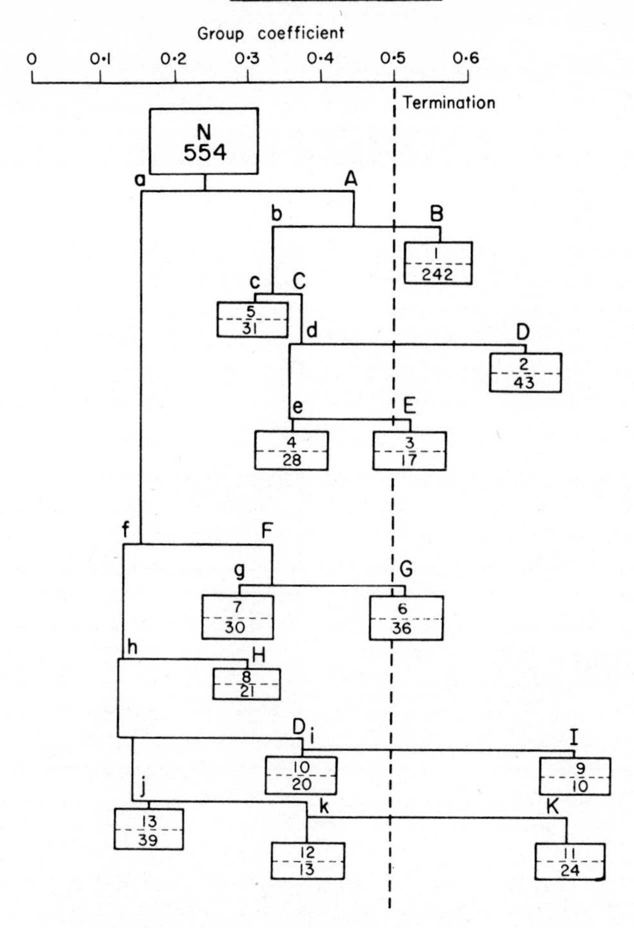

Fig. 1. Hierarchial representation of group analysis based on a survey of 554 quadrats containing a total of 98 species (taken from Crawford and Wishart 1967)

Contouring alone is insufficient to demonstrate the affinities between the terminal groups of the classification, and this is essential to a proper understanding of the relationship between the vegetation and its environment. Basically ordination is an erroneous process as it tries to express in 2 or 3 dimensions what can only properly be explained in *N* dimensions. As Dr Ivimey-Cook (see p.76) has pointed out in his study of the genus *Ononis* the extraction of 12 components is needed to account for 72 per cent of the variation. In ecology, however, the situation is usually simpler and in many studies with temperate vegetation

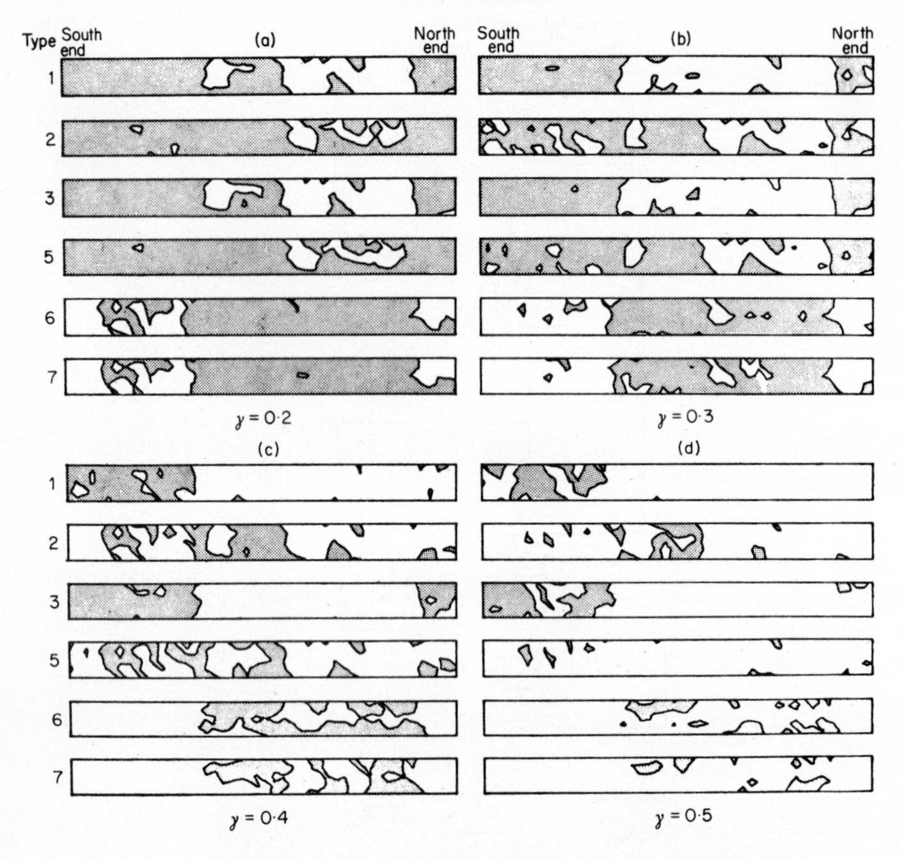

Fig. 2. (a-d) Computer drawn contour maps showing the areas occupied by each of 6 vegetation types when delimited by different levels of (y = level of potential for each quadrat belonging to any particular vegetation type - see Crawford and Wishart, 1968)

(tropical vegetation is especially excluded) two components will account for 75 per cent or more of the variation.

With large surveys (*i.e.* M and N both large) ordination by the procedure of Orloci (1966) and Austin and Orloci (1966) will involve the use of considerable storage space and machine time for the calculation of the correlation matrix, eigenvalues and eigenvectors. The present method (see Crawford and Wishart 1968) replaces the calculation of a matrix for N space by one for K space (where K is the number of terminal groups obtained by using the classification just described).

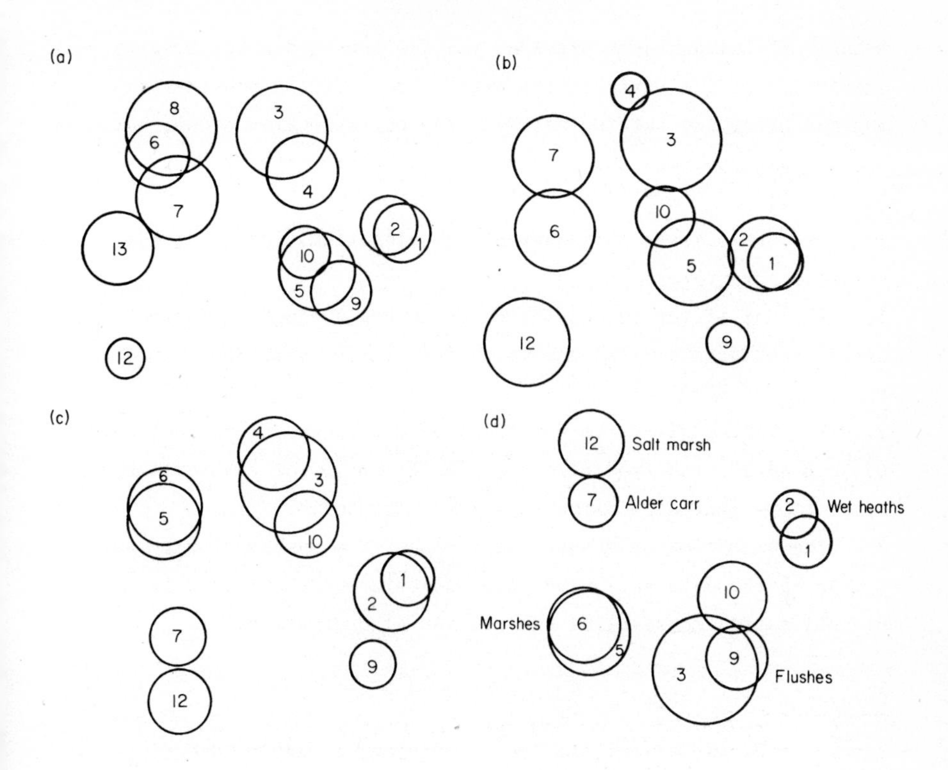

Fig. 3. (a-d) Ordination of groups obtained after agglomeration at different levels of β (β = level of testing for misfits see Crawford and Wishart 1968) (a) β = 0.25; (b) β = 0.35; (c) β = 0.45; (d) β = 0.55

Figure 3 represents the results of this process after the final classification has been checked for possible misfits by an agglomerative procedure (see Crawford and Wishart 1968). Any monothetic divisive classification will give rise to misfit quadrats due to the chance occurrence of a dividing species. The quadrats are, therefore, tested by an iterative procedure, that they are all in their group of best fit. The level at which the fit is considered to be satisfactory can be gradually raised with the effect that can be seen in fig. 3. In this figure a circle is plotted for each group whose centre is the mean position of the group points and radius the standard deviation of the points' radii from the group mean. It follows that provided the two axes represent a large proportion of the variance the size of each circle provides an indication of the heterogeneity of the corresponding group, while the distance between the circles shows the groups' mutual homogeneity.

In summary this approach to classification has three possible advantages:

1. The use of species co-incidence as the ecological parameter avoids a quadratic relationship between survey size and computing time.
2. The retrieval of the potential of every quadrat for belonging to each vegetation type allows the distribution of the vegetation types to be contoured directly by the computer.
3. The reduction of *N* space to *K* space (*K* = number of terminal groups in the data set) uses information already obtained by the classification to simplify the process of ordination.

References

Austin, M.P. and Orloci, L. 1966. Geometric models in ecology II. An Evaluation of some ordination techniques. *J. Ecol.* *54*, 217-27.

Crawford, R.M.M. and Wishart, D. 1967. A rapid multivariate method for the detection and classification of groups of ecologically related species. *J. Ecol.* *55*, 505-24.

Crawford, R.M.M. and Wishart, D. 1968. A rapid classification and ordination method and its application to vegetation mapping. *J. Ecol.* *56*, 385-404.

Orloci, L. 1966. Geometric models in ecology I. The Theory & Application of some ordination methods. *J. Ecol.* *55*, 193-206.

Discussion

Q. (Parks) How long does it take to run a problem with 500 quadrats?

A. 20 minutes on the 1620. All data is on disk storage and much of the time is spent in reading from magnetic disk.

Q. (Barrs)
1. Can you tell us the relative time between this program of division and association analysis?
2. While realising that by representing the clusters as circles you are giving a measure of spread on your "ordination" outputs, could you not show their distribution by outlining the group to indicate its slope.

A.
1. No true comparison available to answer this question, since the programming skills used on the two tasks were not the same.
2. Yes, we could have drawn contours but we felt this was a clearer method.

Q. (Jeffers)
1. Have you investigated the use of the results of such an analysis from a systematic or partly random, partly systematic sample in combination with trend surface plotting to aid the interpretation.
2. Do you visualize the extension of these techniques to cover repeated surveys of the same area in time, so that changes in the ecology of an area can be detected and mapped?

A.
1. We have taken the final groups and contoured these using a trend surface program developed by Dr Cole at St. Andrews.
2. Yes, the process can be repeated later to do this. Several surveys taken at different times have been stored so that later surveys may be compared with them.

Comment: (Read) The Chairman Dr Jeffers has asked if the technique for using three dimensional trend surfaces had been worked out. Such a program dealing with three geographical variables plus one geological variable has been published in the following paper: Harbaugh, J.W. 1964. A computer method for four-variable trend surface analysis illustrated by a study of oil gravity variations in South-eastern Kansas. *Kansas Geol. Survey Bulletin, 171*, 58 p.

Q. (Ross) Could Mr Wishart please explain what is meant by ordination, and how his diagrams were actually obtained?

A. (Wishart) The original technique was to impose on the diagram of the principal components plane some other factors, such as Ph, in a suitable coding scheme; but in fact we don't use this. By ordination we mean the plotting of the cluster points in the principal plane in a summarized form with a circle of a given radius and a given variance. (All this is in the paper).

Comment: (Sneath) It may be worth noting that although the term "ordination" has been used in various special senses, it is a useful word for any method that arrays the entities under study along a scale or scales, without dividing them into sharp groups or classes. in this wide sense, which is now quite commonly used, it contrasts with cluster methods and the like, which construct distinct classes of entities.

Q. (Hall) Is it felt that the use of presence/absence data is satisfactory in this work? I have experience of this in a salt marsh where gradients were shown by abundances but many of the species were present throughout.

A. Had we been doing details of any one of the slacks, we should have required this information, but for a rapid survey the presence/absence method was preferred for simplicity.

A NUMERICAL TAXONOMY ON BUSINESS ENTERPRISES*

Friedhelm Goronzy

4620 Castrop-Rauxel, Gerther Str. 51, Germany

Management and Organization theorists are becoming increasingly concerned with the limitations of their theories and findings. Such developments are a sign of the growing awareness of the problems of methodology in organizational research. More and more students argue for a recognition of boundary conditions of the various research results reported in the literature. These tendencies seem to indicate a search for specific theories with considerable predictive power and corresponding systematic classification frameworks for organizational phenomena. It is the purpose of this paper to report an attempt to classify business enterprises on the basis of several measurable characteristics.

Data Collection

The data for this study was collected through a rather comprehensive mail questionnaire which was distributed to 500 American manufacturing firms. Because of the rather lengthy questionnaire and the confidential nature of much of the information requested, only 58 firms submitted usable replies and of these respondents only half revealed their identity. The analysis itself is based on 50 replies which were received prior to the deadline.

* This paper is based in part on Appendix B of Friedhelm Goronzy, "A Multivariate Analysis of Selected Variables of Manufacturing Business Enterprises", Unpublished Dissertation, Louisiana State University, 1968. The writer acknowledges the support of this work through a dissertation year fellowship granted by the Graduate School of Louisiana State University. The research was supervised by Professor Herbert G. Hicks.

Methods of Analysis

The collected information was subjected to various multivariate analyses in order to gain a better understanding about the behaviour of the variables, *e.g.* principal component analysis with subsequent varimax rotation on the original and logarithmically transformed variables. The result of these analyses is reported elsewhere in greater detail.[1] The cluster analysis of the 50 manufacturing enterprises was performed with the NT-SYS computer programs developed by Professor Sokal and his associates at the University of Kansas.* Of the various analyses two dendrograms are reproduced here. Fig. 1 depicts the classification on the basis of correlation coefficients between enterprises or taxa and Fig. 2 is based on the taxonomic distance. In both cases the most meaningful classification was obtained by average linkage. Single linkage dendrograms exhibited the familiar chaining effect and were not investigated in great detail.

Presentation of Findings

Although there may be serious objections to a classification on the basis of correlation coefficients as shown in Fig. 1, this method, nevertheless, produced very meaningful results as will be shown below. The classification in Fig. 2 based on taxonomic distances is included only for comparison purposes for the reader and will not be analysed further. The discussion is to be based on Fig. 1.

* The writer is greatly indebted to Professor Robert R. Sokal and Mr Ronald Bartcher for analysing the data with the NT-SYS computer programs. For a write-up of the programs see F. James Rohlf, John Kishpaugh and Ron Bartcher, "Numerical Taxonomy System of Multivariate Statistical Programs", The University of Kansas, Lawrence, Kansas, July, 1967. Preliminary analyses of the data were performed with a single linkage cluster analysis computer program developed by Miss Judith Fiehler of the Louisiana State University School of Music. (See Appendix).

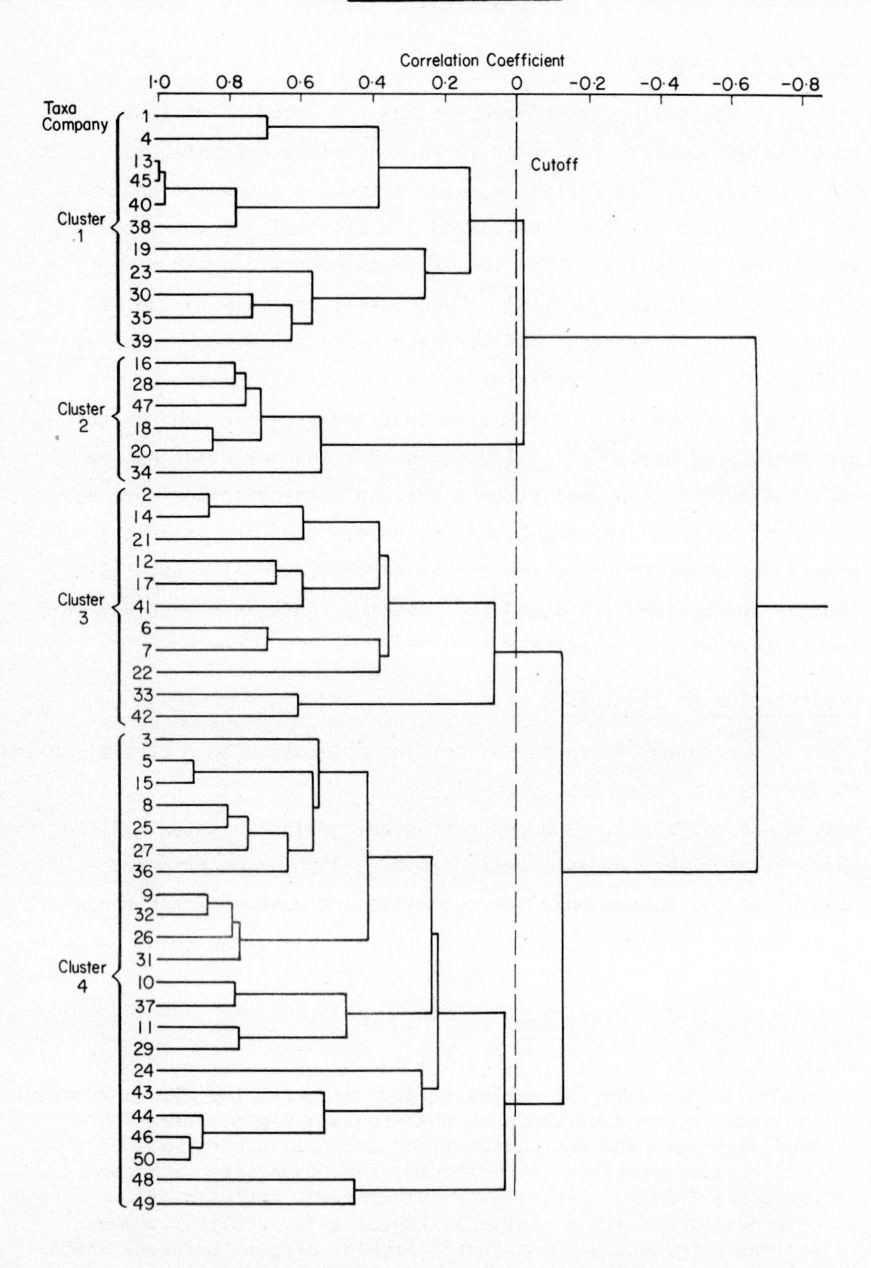

Fig. 1. Dendrogram based on correlation coefficients joined by average linkage

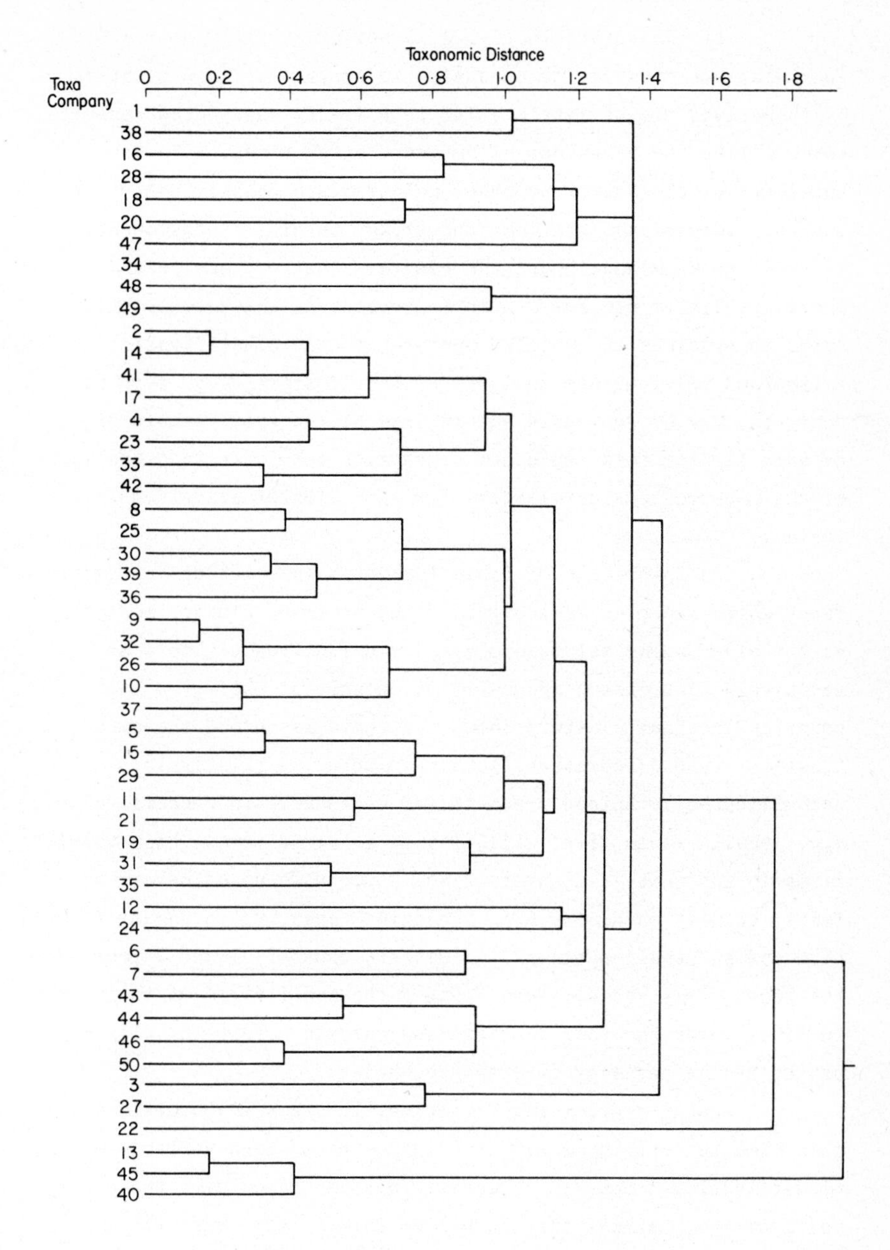

Fig. 2. Dendrogram based on taxonomic distance coefficients joined by average linkage

If the cutoff in Fig. 1 is set arbitrarily at $r = 0.0$ four distinct clusters of unequal size emerge. These clusters by themselves are of little value in a fairly unexplored research subject like the behaviour of business enterprises. To the knowledge of the writer no other multivariate classification of business enterprises has been undertaken before. Consequently there is no knowledge about the clusters that may emerge and therefore little use for *a priori* hypotheses which could only serve to preclude an unbiased approach to the classification results. A rigorous multivariate analysis of the clusters, *e.g.* component analysis, may be desirable but in view of the small sample size of each cluster this could not be carried out. Instead the means of the measured characteristics for each cluster are given in Table 1.

An inspection of Table 1 reveals some rather interesting facts which are also indicative of the power of cluster analysis as a tool of numerical taxonomy. By definition clusters are relatively homogeneous groups. An anlysis of the firms that comprise the four clusters shows that this is indeed the case. Clusters 1 and 3 represent firms manufacturing relatively uncomplicated technical products and components in greater volume, *e.g.* fans. These firms will also be referred to as "high volume" firms in the text. Clusters 2 and 4 are made up of makers of fairly complex machinery *e.g.* , machine tools, which appear to be produced in "small batches". Further, clusters 1 and 2 represent the large firms and clusters 3 and 4 the small firms of this survey. In other words the four clusters approximate a four-way classification on the basis of size and technology.

Being a pilot investigation, it seems inappropriate at this time to "overinterpret" the influence of each variable on the classification process. However, in conjunction with the other multivariate analyses that have been carried out (but will not be reported here) a brief account of the nature of each variable can be given. This discussion will take up the remainder of this paper.

Variables		Cluster 1 11 Firms	Cluster 2 6 Firms	Cluster 3 11 Firms	Cluster 4 22 Firms
Total Sales	1	$21,400,000	$53,200,000	$ 8,400,000	$10,000,000
Direct Sales	2	$ 6,500,000	$47,600,000	$ 7,000,000	$ 5,800,000
Other Sales	3	$14,900,000	$ 5,600,000	$ 1,400,000	$ 4,200,000
Sales per Employee	4	$ 20,250	$ 25,880	$ 19,400	$ 29,240
Sales per Production Worker	5	$ 32,000	$ 50,680	$ 36,100	$ 77,400
Customer Accounts	6	1,920	2,600	5,740	757
Customer Orders per Month	7	800	630	2,040	854
Order Changes per Month	8	15	94	153	58
Total Assets	9	$17,900,000	$38,400,000	$ 5,300,000	$ 6,400,000
Fixed Assets	10	$ 4,600,000	$10,000,000	$ 1,500,000	$ 1,700,000
Capital-Output Ratio	11	0.67	0.72	0.57	0.69
Technology-Capacity Index	12	18,750	52	40,470	930
Parts Orders per Month	13	430	623	892	612
New Products in 3 Years	14	8	5	5	10
Average R & D Time (Years)	15	1.7	1.6	0.9	1.4
Engineering Changes	16	58	375	89	32
Part Numbers	17	8,250	30,500	5,285	9,872
Total Employment	18	986	2,070	415	358
Production Workers	19	600	1,055	225	183
Number of Foremen	20	41	56	15	7
Number of Other Superiors	21	58	124	38	37
Unit of Supervision	22	15	23	13	27
Number of Subordinates	23	4	7	4	3
Division of Labor	24	0.96	0.98	0.92	0.8
Sales Department	25	58	157	65	55
Manufacturing Department	26	755	1,517	291	243
R & D Department	27	40	196	20	29
General Administration	28	55	204	40	36
Fixed Assets/Production Worker	29	$ 5,030	$ 8,950	$ 6,420	$ 15,025

Table 1. Means of 29 Selected Variables for 4 Clusters Joined with Average Linkage on the Basis of Correlation Coefficients

The variables Total Sales (annual), Total Assets, Fixed Assets and Total Employment can be viewed as measures of size. These variables have very high intercorrelations and the highest factor loadings on the first component. In further analyses it seems appropriate to reduce the number of size variables in order to limit the extent of bias or artificial weighting relative to size. Of course, this entails also the possibility that other information may be lost by the elimination of certain variables.

Direct Sales refers to sales through company salesmen whereas Other Sales indicates sales through middlemen and other marketing channels. The tendency of makers of complex machinery to sell through their own sales force is most pronounced in cluster 2. A high correlation of Direct Sales and Engineering Changes (processed per month) especially in this cluster suggest a possible information feedback relationship. This information feedback seems to occur in addition to the fact that complex machinery usually needs more development work than less complex products. The existence of such a relationship is also suggested in a letter of a company president to the writer. This man stated that his sales force had been re-organized for that very reason.

The variable Sales per Employee, Sales per Production Worker and Capital-Output Ratio (Total Assets/Annual Sales Volume) can be viewed (with some caution) as measures of efficiency of firms within a cluster. There are substantial differences in the value of these variables between firms in the same cluster. This suggests varying degrees of efficiency from firm to firm in the same line of business. However, for an intercluster comparison value-added information would be more appropriate. Unfortunately, this measure is not available.

Variables that are indicative of the distribution pattern of the goods manufactured are Customer Accounts, Customer Orders per Month and Order Changes per month. As Table 1 shows these variables measure this pattern fairly well. The small batch manufacturers in cluster 2 have fewer customers and orders than the high volume producers in cluster 3. This relationship holds in spite of the fact that the firms in clusters 2 are much larger than those in cluster 3. The variable Parts Orders per Month was inserted in order to measure the amount of machine business versus parts business but the desired differentiation did not show up in the analysis.

The variable Technology-Capacity Index (TC Index) is an attempt to quantify the concept of production capacity in relation to the technological level of the product. Most of the surveyed firms had multiple product lines and it is therefore difficult to speak of capacity without specifying the product line. Several attempts to construct a meaningful index of capacity failed. Of all the measures tried the simple geometric mean gave the best approximation of the concept of capacity. For example, if a firm reported the following production figures: 27 units per month, 100 units per month and 10,000 units per month then the Technology-Capacity Index is calculated by $\sqrt[3]{27 \cdot 100 \cdot 10{,}000} = 300$. The same procedure was followed for 2, 4 or more produce lines. The behaviour of this TC Index is reasonably well in line with other measures that vary primarily as a result of technological considerations such as Engineering Changes and Part Numbers.

Part Numbers refers to the number of parts kept in inventory for customer service purposes in the widest sense. As can be seen the makers of complex products - clusters 2 and 4 have to keep a larger inventory of these parts than the high volume manufacturers. Other variables which were deliberately inserted to measure the interaction between the Research and Development (R & D) effort and the level of technological complexity of the products did not always give the anticipated results. For example, New Products in 3 Years refers to the number of new products brought out within the last three years. Average R & D Time (Years) measures the average length of the development time from the decision to build a prototype to the decision to put the Product on the market. It was anticipated that fewer complex products would be marketed in the same time span and it would take considerably longer to develop them. As the table shows the results do not show this simple relationship. It should be added that the response to these questions probably has been influenced by judgmental considerations and therefore the answers may not be strictly comparable. However, in the case of variable 27 - the number of people in the R & D Department - it can be seen that the makers of machinery in cluster 2 employ decidedly more R & D personal (absolutely and relative to size) than the firms in the other clusters.

The relationship between the variables Production Workers, Number of Foremen, Number of Other Superiors, Number of Subordinates and the number of personnel in the Sales and Manufacturing Departments and in General Administration seems to indicate that larger firms enjoy economies of scale. However, the relationship does not always emerge very clearly probably due to the fact that the influences of technology and size do interact. Whether the administrative overhead is constant or increases or decreases with size is not only a matter of counting the number of people and their titles but much more so a matter of ascertaining what they actually do. Although it seems that the small firms have fewer subordinates per superior it is probably that the supervisor in the small company does part of the actual work in addition to his supervisory task. The difficulty of distinguishing clearly the influence of the various forces in an enterprise is just another indication of the great help that the judicious application of multivariate methods may provide to the researcher for studying the behaviour of business enterprises.

The Unit of Supervision is the size of the work group supervised by a first-line foreman. As the table shows clusters 2 and 4 comprising manufacturers of machinery have higher units of supervision than the high volume firms in clusters 1 and 3. This can be attributed to the fact that heavy machinery is generally produced by unit centered production methods. A group of highly skilled people assembles a machine. High volume production most often is workflow oriented and often requires less skilled labour which may need more supervision and therefore lead to smaller work units.

The concept of Division of Labour has been adapted from other sources and is more fully discussed elsewhere.[1] It is based on the assumption that units of supervision (Z) in production are relatively homogeneous with respect to the crafts or trades, *e.g.* lathe operators, assemblers, etc. will form fairly homogeneous units of supervision. If this homogeneity assumption is reasonable then division of labour is relatively pronounced if the number of units of supervision is large in relation to the total production workforce (X) or division of labour $D = 1 - \frac{\Sigma(Z)^2}{X^2}$. A value close to 1 signifies a great degree of division of labour. Consequently, in cluster 4 where the firms are relatively small and the units of supervision are large, the extent of division of labour is small. On the other hand, cluster 3 shows that division of labour increases even in small firms if the size of the unit of supervision declines. The fact that the larger firms in cluster 1 and 2 show a high division of labour merely indicates that large firms have greater degrees of division of labour regardless of technological considerations.

Finally, the variable Fixed Assets/Production Worker seems to illustrate two points. First, machinery manufacturers employ relatively more fixed capital per production worker than the high volume manufacturers of this survey. This is somewhat contrary to the expectations before the survey. Second, the small firms employ relatively more fixed assets per production worker than the larger firms. In the case of the small machinery manufacturers in cluster 4 this relationship is especially clearly illustrated and may be an example of the ill effects of Babbage's "principle of the least common multiple" on the economy of the small firm. That is to say that the small firm probably has to invest disproportionately more in fixed plant and may be unable to utilise it as effectively as the large firm.

Thus on the basis of the evidence in Table 1 it appears that Numerical Taxonomy is of great value for the analysis of the business enterprise. However, because of the small sample size of each cluster and the relatively greater influence of missing data on the results the analysis cannot be extended beyond a simple comparison of means.

Further studies in Numerical Taxonomy of business enterprises are needed to confirm the findings of this pilot study and refine the interpretations. Also, the variables to be employed in other studies should measure a wider range of phenomena (including behavioural) than those used in this investigation and some of the redundant variables should be deleted. An investigation of the effect of derived variables (ratios, geometric means) on the classification process is also needed.

Reference

1. Friedhelm Goronzy, "A Multivariate Analysis of Selected Variables of Manufacturing Business Enterprises", Unpublished Dissertation, Louisana State University, 1968.

Discussion

Comment: (Jeffers) He expressed his doubts about the manner in which the diverse characteristics concerning the businesses were used in the classification.

A. This was exactly what the author wished to improve.

GROUP FORMING AND DISCRIMINATION WITH HOMOGENEITY FUNCTIONS

A. V. Hall

Bolus Herbarium, University of Cape Town, Rondebosch,
C.P., South Africa

Numerical taxonomy is most often taken to refer to the problem of forming groups that are optimal with respect to certain criteria. This is only one operation among many in the subject of taxonomy, a meeting point of many skills. A fully-fledged taxonomist must not only know about plants: he must also be linguist, historian, geographer, ecologist and often, collector in remote places. Although numerical aids can be used with profit in more aspects of taxonomy than group-forming, by far the greater part is left to the abilities of the taxonomist. With this in mind, 'numerical aids to taxonomy' gives a less ambitious description and offers a reminder of the variety of methods that are being developed under this heading. The methods may be generalised and grouped as follows:

1. Group-forming: (a) Lowest level using specimens from the field. (b) Higher levels using the abstractions set up in (a).

2. Discrimination and identification: (a) Matching un-named specimens with groups that have been set up previously using weighted characters where necessary. (b) Finding the best position of new taxa in previously described arrangements. (Strictly, the entire arrangement of a group should be reviewed on entry of a new member, but this is scarcely practical with present systems of study and record keeping). (c) Key-forming, or finding the best way of splitting up a group into equal-as-possible sets on the basis of the most easily observed properties, and allowing later subdivisions also to be optimal in these respects.[1]

3. Finding typical and atypical taxa in a group: Showing which taxa have on average (1) the least peculiar and (2) the most unusual properties in a given group.[2,3]

4. Showing geographical trends of variation.[4]

5. Phylogeny: Finding the sequences with least change from taxon to taxon, and setting up the branching sequence called a cladogram.[5]

This paper is confined to group-forming and identification. Both these terms have been confused under the word *classification*. This has been unfortunate as the data about the properties of taxa and specimens are used differently in the two procedures. ('Classification' is best reserved for a pynonym of 'arrangements of items').

The groups with which specimens or taxa are to be identified are each more or less homogeneous for a number of their properties. These properties will get first consideration in setting up identification keys. Another criterion for this 'weighting' is ease of observation.

In group-forming however, the aim is generally to place the taxa or specimens into sets that are as homogeneous as possible for as many attributes as possible. In a sense we are setting up an information retrieval system that must be as useful for as many properties of plants that we can find. Later, if a specimen is identified with a group, one should be able to predict it probably has many of the other properties of the group that one has not observed in the specimen. Some biological qualifications may be added to this, but the essentials of the dictum seem to stand well in relation to past practice. In this approach there may be some properties that are almost superfluous and because of being nearly homogeneous throughout the group have little influence on the ways the linkages are formed. As such they may be under-weighted, and it is important to note that in this way the differing influences of the properties occur as an *integral part* of the grouping procedure.

The next problem is a major one and does not seem to have had the attention it deserves. Are the items being grouped *samples* from populations of things or are they *unique* and form a closed array which is unlikely to change during the useful life of the classification?

Using homogeneity analysis different methods must be used for the two situations. This is shown in the example given in

Table 1 and Figs. 1(a) and 1(b). The groupings which are as homogeneous as possible for this data (according to the formula

Item number:	1	2	3	4	5	6	7
Attribute values for the only property :	1	1	1	2	2	2	3

Table 1. Data to illustrate the unique items/samples problem

discussed below), are given in the dendrogram in Fig. 1(a). Each linkage is the result of finding the most homogeneous of all the possible trial groupings at each stage. The homogeneity is calculated using all the data for each character row in a sub-group. Where a sub-group is already large, the inclusion of a slightly anomalous item may cause little disturbance to its overall heterogeneity. There is a definite preference for item 7 to attach to group 4 - 5 - 6 before group 1 - 2 - 3. Group 1 - 2 - 3, although as distant on average from 4 - 5 - 6 as item 7, consists of three members and causes a larger disturbance in heterogeneity.

If the seven items were unique, this would be the 'tidiest' arrangement. If however, they were samples forming an open array, it would seem that item 7 comes from a population that is under-represented in the data. It is then not appropriate that there should be a preference for linking item 7 to group 4 - 5 - 6. The way of avoiding this is to regard each sample as representing a *class*. Also, each class must be equally represented in trial fusions, restricting the homogeneity calculations to two items at a time. The test for fusion is then: how homogeneous is the class-of-such-things as item *A* grouped with the class-of-such-things as item *B*?

More than two-membered clusters are then it seems best formed by finding how homogeneous on average are all the possible pair-groups that can be formed between the class-of-such-things as the items of cluster *A* and those of *B*. This is the same as the

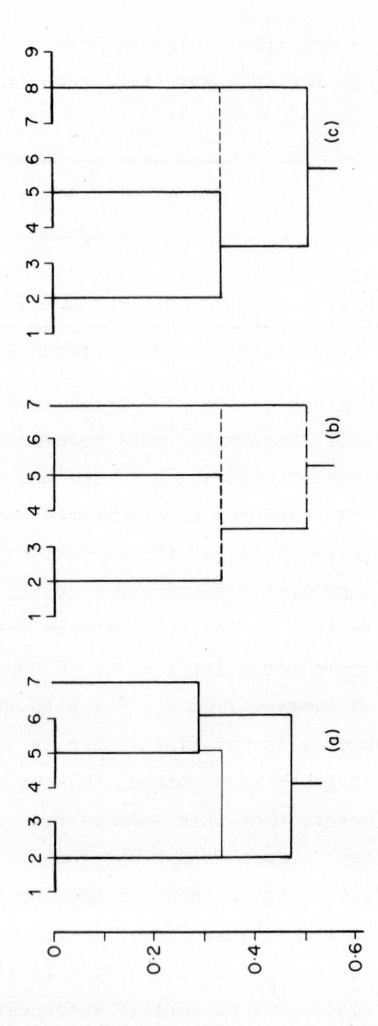

Fig. 1. Dendrograms showing the grouping of the items given in Table 1, with (a) full heterogeneity grouping and (b) average linkage with heterogeneity computed for two membered groups only. In (c) both methods were used and two items exactly like No. 7 were included. Average linkage levels are shown dashed, alternatives in thinner lines. Vertical scale is in heterogeneity units from 0 to 1.

unweighted pair-group average linkage system.[6] For the data in Table 1, the method gives an appropriate ambiguity between the linkage of group 4 - 5 - 6 to either item 7 or group 1 - 2 - 3. (See Fig. 1(b)). Increasing the number of items identical to 7 (giving better representation) has no effect on the average linkage level or the general sequence of linkings. (See Fig. 1(c)).

The levels formed in this way give an approximate idea of group distinctness. Once the groups are formed, more precise studies of differences may be made using the changes in heterogeneity that occur when equal numbers of samples from two groups are fused. This is described later.

Heterogeneity Functions

Heterogeneity functions have been developed for two-state and quantitative data. There is also a function for ecological data that gives emphasis to the results from the taxa with the higher average abundances in the plots of each trial group.[7,8]

The two-state function is designed for non-modal, presence/absence data coded as presence = 1, absence = 0 or vice versa. It is written as

$$H_t = \frac{1}{p} \sum_{j=1}^{p} \frac{n_{ajr}}{n_{ajrh}}$$

where for the jth of the p properties, the actual number of rare states in the trial group n_{ajr} is divided by the number of rare states for an imaginary, maximally heterogeneous group of the same size, n_{ajrh}. Some results obtained with the function are shown in Table 2.

The heterogeneity function for quantitative data reads as follows:

$$H_q = \frac{1}{p} \sum_{j=1}^{p} \frac{s_{jn}}{s_{jnh}}$$

where s_{jn} is the standard deviation of the row of values for a property in a set of objects and s_{jnh} is the same for an imaginary

Data	n_{ajr}	Data for imaginary heterogeneous group	n_{arjh}	H_t value
1 1	0	0 1	1	0
1 0	1	0 1	1	1
0 0	0	0 1	1	0
0 0 0	0	0 0 1	1	0
0 0 1	1	0 0 1	1	1
0 1 1	1	0 0 1	1	1
0 0 0 1	1	0 0 1 1	2	$\frac{1}{2}$
0 0 1 1	2	0 0 1 1	2	1
0 1 1 1	1	0 0 1 1	2	$\frac{1}{2}$

Table 2. Showing the distribution of the values of the Heterogeneity Function H_t for sets of two-state data.

set with maximal heterogeneity. Examples showing the distribution of the function are given in Table 3.

Data (scale maximum 1)	Data for imaginary heterogeneous group	H_q value
0.3 0.7	0 1	0.40
0.3 0.4 0.6 0.7	0 0 1 1	0.31
0.3 0.4 0.5 0.6 0.7	0 0 1 1 1	0.29
0.4 0.6	0 1	0.20
0.4 0.5 0.6	0 1 1	0.17
0.8 0.9 1.0	0 0 1	0.17
0.4 0.5	0 1	0.10
0.0 1.0	0 1	1.00
0.3 0.3	0 1	0.00

Table 3. Showing the distribution of the values of the Heterogeneity Function H_q for sets of quantitative data.

The function reduces to a simple modulus for two-membered groups. This can be shown for a property j that has been scaled to a range with a maximum of 1, as follows:

$$H_{qj} = \sqrt{\frac{a_{jx}^2 + a_{jy}^2 - \frac{1}{2}(a_{jx} + a_{jy})^2}{1^2 + 0^2 - \frac{1}{2}(1 + 0)^2}}$$

$$= \sqrt{\frac{\frac{1}{2}(a_{jx}^2 - 2a_{jx}a_{jy} + a_{jy}^2)}{\frac{1}{2}}}$$

$$= \pm(a_{jx} - a_{jy})\text{ , or more generally,}$$

$$H_q = \frac{1}{p}\sum_{j=1}^{p} |a_{jx} - a_{jy}|$$

This function resembles, among others, the Mean Character Difference of Cain and Harrison,[9] derived on different grounds. For two-state data and two-membered groups, the H_t function converted for Homogeneity $(1-H_t)$ becomes identical to the Simple Matching Coefficient of Sokal and Michener.[10] The H_q function for two-membered groups is preferable to Taxonomic Distance,[6] where the larger differences are over-emphasised in the process of being squared before addition. This is shown for some sets in Table 4. It may be noted that the differences between the results are not exceptionally large, and Taxonomic Distance has given helpful results in a number of studies. However, although its semi-geometric derivation is attractive, Taxonomic Distance has distortions that make it clearly not suitable. This view has also been put forward by Colless.[11]

The Homogeneity Function for grouping sites or vegetation types in ecology has a modulation system for giving greater emphasis to the values from the more common taxa in a trial set. A rare species at the same abundance at two sites contributes less to the overall homogeneity value than a common species. The function is

Property Number	Data for two objects, scale maximum = 1.		$H_q = \frac{1}{p} \sum_{j=1}^{p} \lvert a_{jx} - a_{jy} \rvert$	$d = \sqrt{\frac{1}{p} \sum_{j=1}^{p} (a_{jx} - a_{jy})^2}$
1	0.1	1.0	$\frac{1}{2}(0.9 + 0.1)$	$\sqrt{\frac{1}{2}(0.81 + 0.01)}$
2	0.1	0.2	$= 0.50$	$= 0.64$
1	0.1	0.5	$\frac{1}{2}(0.4 + 0.1)$	$\sqrt{\frac{1}{2}(0.16 + 0.01)}$
2	0.1	0.2	$= 0.25$	$= 0.29$
1	0.1	0.3	$\frac{1}{2}(0.2 + 0.1)$	$\sqrt{\frac{1}{2}(0.04 + 0.01)}$
2	0.1	0.2	$= 0.15$	$= 0.16$
1	0.1	0.2	$\frac{1}{2}(0.1 + 0.1)$	$\sqrt{\frac{1}{2}(0.01 + 0.01)}$
2	0.1	0.2	$= 0.10$	$= 0.10$
1	0.1	1.0		
2	0.1	0.2	$\frac{1}{2}(0.9 + 0.1 + 0.1)$	$\sqrt{\frac{1}{2}(0.81 + 0.01 + 0.01)}$
3	0.1	0.2	$= 0.37$	$= 0.53$

Table 4. Showing comparison of results for the Heterogeneity Function H_q and the Taxonomic Distance Function d, using sets of data with various differences for each property.

written for a subset of k plots as follows, and its behaviour in a trial grouping of two plots is shown in Table 5.

$$H_{qm} = \sum_{j=1}^{p} \left[\left(\sum_{t=1}^{k} a_{jt} \right) \left(\sum_{j=1}^{p} \sum_{t=1}^{k} a_{jt} \right)^{-1} \right] \left(1 - \frac{s_{jk}}{s_{jkh}} \right)$$

Species	Abundance data for species on 2 plots, scale maximum = 1.		Modulating Factor, M	Homogeneity value, H	Product $M.H.$
1	0.0	0.0	0/5.8=0	1	0
2	1.0	1.0	2/5.8=0.35	1	0.35
3	0.1	0.1	0.2/5.8=0.04	1	0.04
4	0.6	0.4	1/5.8=0.17	0.8	0.14
5	0.0	1.0	1/5.8=0.17	0	0
6	0.8	0.8	1.6/5.8=0.28	1	0.28
			H_{qm} value for trial group = 0.81		

Table 5. Showing the behaviour of the Homogeneity Function for ecology, H_{qm}, in a trial grouping of two plots with abundance data recorded for six species.

Coding of Properties

It would seem that more work is needed on the philosophy of property coding. Small changes in approach can have important effects and it is essential that optimal criteria be laid down for general guidance. This work is certainly as important as the finding of appropriate functions for computation.

Both two-state and quantitative data may be used together in the same study using a compound heterogeneity function:

$$H_{tq} = H_t + H_q$$

Presence or absence properties may be coded simply as 0/1 data for the two-state function. Pure two-state properties seem to be rarer in biological material than in fields such as linguistics and archaeological artifact studies.

Shades of variation should be given if possible, as scores or measurements of lengths, shapes, angles or weights. Each row of values for a property is scaled to a range with a maximum of 1 for convenient use of the quantitative heterogeneity function. This is done by dividing all the values in a row by the largest. No matter how large the absolute differences are, small proportional differences give small changes in heterogeneity. There may be some objection to using the largest value as a reference point. Such values may come from abnormal objects, and they have an important influence on the contributions (weightings) of the various properties. For the moment this seems to be a logical necessity but further study is needed on this. So far there has been no need to record a property in non-ordered multi-state form for heterogeneity analysis.

With some qualifications, the inclusion of a property may be based on its varying in at least a slightly different way from the others. This rule may not be valid where there is strong co-variation between functionally different properties, all of which should then be included.

Where there is a certain degree of co-variation among a set of properties, one should avoid giving absolute sizes (measurements or direct scores) to more than one of the set. One might otherwise have the smaller forms of a set of taxa cluster together even though the relative proportions of their structures differ. Such properties are best given as ratios of their size relative to the one for which the measurements are given. It is a problem to know at what level of co-variation this rule should be used. Should there be measurements for leaf, stem, bract and flower size? To the purist, all the measurement data should be taken as independant items of information until the basis of the co-variation has been found. To the taxonomist, the impression of linkage of properties suggests that giving all their measurements would mask the effects of the other features, being poor for information retrieval.

Among the many kinds of properties, chemical data seem to be the most difficult to score. With details of synthetic pathways known to us the problem may be largely resolved by coding the degree of departure from a starting material in, for example, the growth of side-chains. Each development might be coded separately and the heterogeneity results averaged for each given class of compound, this value entering the set of heterogeneity values from other properties. Without quite complete data on pathways this scheme may become unreliable. The alternative of coding the presence or absence of compounds as two-state data may impose severe biases that cannot be justified.

This example of chemical features mirrors much of our ignorance of the basis for coding shapes and sizes, which are the result of morphogenetic processes that are so poorly understood. If numerical methods serve only as a spur to work on morphogenesis, they will have made a most important contribution.

Identification

Identification problems are large in areas where the flora is rich, and where there are not enough local specialists for adequate coverage of critical groups. Keys are helpful if the specimens have most of the parts that are quoted. They can be most inadequate for naming non-flowering material from, for example, regeneration studies or in areas where the flowering season is short. Problems of this kind are most acute in the Southern Cape Province in South Africa, where there is a flora of the order of some five thousand species of Phanerogams alone.

With this in mind a program has been written to explore the possibilities of computer-based identification. Data from an unknown specimen is compared with that in a more or less complete reference matrix set up in the computer beforehand. Where one of a prior chosen set of characters disagrees beyond a selected level the possibility of a match is rejected and the next comparison is started immediately. The comparisons are based on the average homogeneity over all properties of two-membered test groups. Thus although the method is basically polythetic, there is a monothetic control for rejecting impossible choices, speeding up each run with the program. The output gives a record of the ten best matches for the unknown, and, if desired, the characters that disagree most so that they may be checked in the Herbarium if necessary.

A problem of special importance to palaeontologists is the identification of material in groups separated by slight quantitative differences. Fisher's discriminant analysis, first described in 1936, has been used as an aid in this work. The measurements are multiplied by factors which are adjusted statistically to give maximum separation of pairs of groups when the products are added together:

$$z = w_1m_1 + w_2m_2 + w_3m_3 \ldots w_nm_n,$$ where

$w_1 . . w_n$ are the weighting factors,

$m_1 . . m_n$ are the measurement values for the n characters for each specimen.

The calculation of the weightings can be tedious for more than six or eight characters and very lengthy indeed for fifteen or twenty. There is a more rapid way using heterogeneity analysis in which large numbers of characters may be included with relative ease. Like discriminant analysis, the method is used for two groups at a time. Each property has a particular heterogeneity level in the two groups. On fusion this generally changes to a new level. This is shown in a simple example in Table 6. For some properties the average increase in

Attribute values for a property in two 2-membered groups :		0.7 0.8	0.9 1.0
Heterogeneity values :	Before fusion:	0.10	0.10
	After fusion:	0.22	

Table 6. Showing the change in Heterogeneity of a property on the fusion of two groups of objects.

heterogeneity on fusion will be larger than usual; these are the characters that differ most between the two groups. The amount of increase in Heterogeneity gives a measure of the amount of weighting for each quantitative property in a discriminant function. The sign of the weighting is adjusted according to whether one group has larger values for a property than the other. Further details of the method are given elsewhere, including a simple routine for non-modal, two-state data.[13]

On the basis of 49 of the 50 specimens' sets of measurements given by Fisher,[12] the discriminant function for the characters of *Iris setosa* and *I. versicolor* becomes

$$Z = 0.00853m_1 - 0.001379m_2 + 0.008632m_3 + 0.025374m_4 ,$$

where m_1 is the sepal length, m_2 the sepal width, m_3 the petal length and m_4 the petal width. This gives a distribution that is generally similar to Fisher's results. (See Fig. 2).

Where a property is entirely different in two groups, the heterogeneity change on fusion of equal subsets is 1 and it may be said that there is a character difference of 1. Smaller changes may be recorded in other features, and all such values may be totalled to give a figure of the equivalent overall character difference. This is interesting for the study of relative rank as one might say, for example, that species should not differ from the nearest similar form by more than 2.0 character difference equivalents. To preserve the validity of this the properties should be as independant as one may wish to have in biological terms or for information retrieval. Careful observation of criteria for property coding will be needed for getting comparable results.

This conclusion applies in so many other aspects of this work. With better numerical aids becoming available, it is time to call for more research in the coding of the raw data.

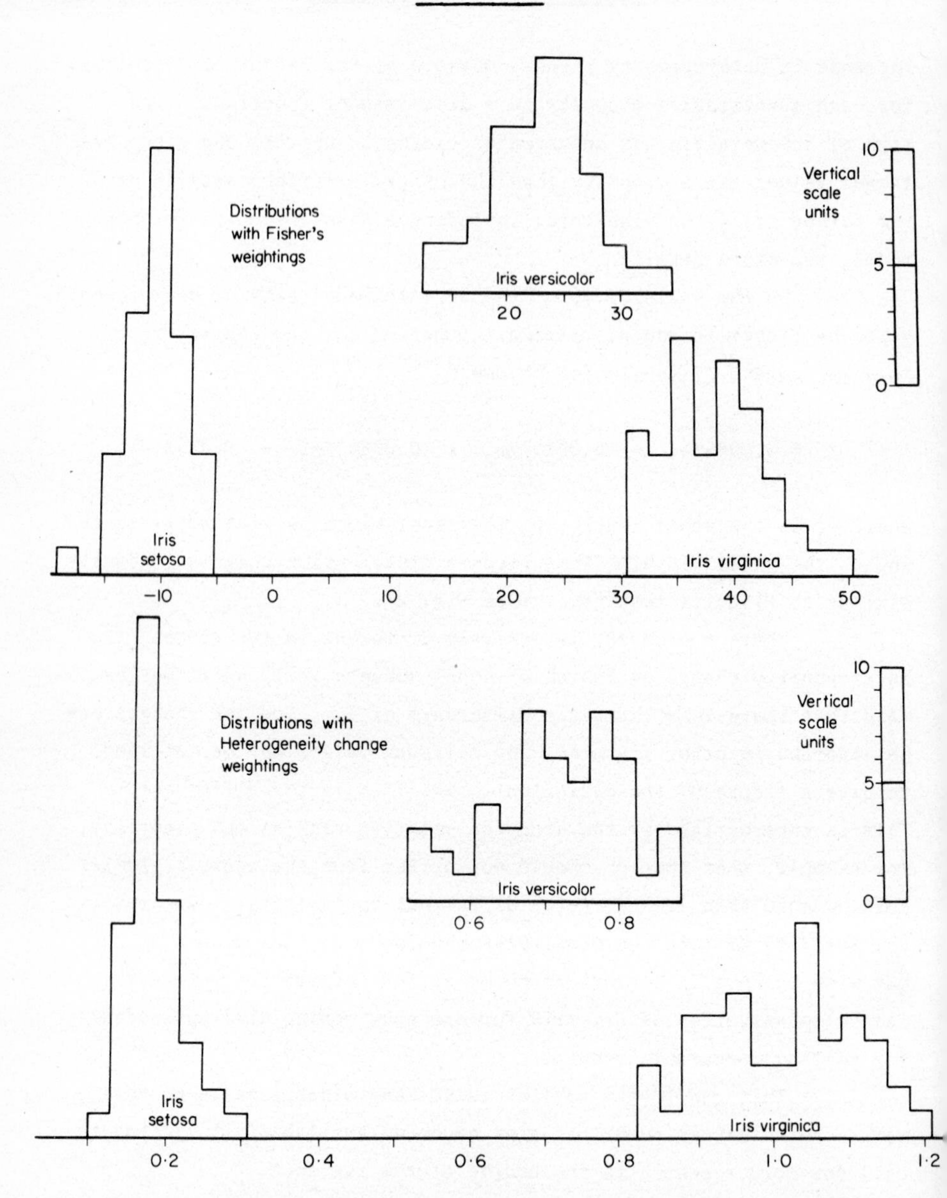

Fig. 2. Showing the distributions of 50 specimens each of Iris setosa, I. versicolor and I. virginica, using a linear compound of weighted measurements of sepal length and width and petal length and width, the weightings being calculated by Fisher's method and by the Heterogeneity change method.

References

1. Osborne, D.V. (1963): Some aspects of the theory of Dichotomous Keys. *New Phytol.*, *62* : 144-160.
2. Hall, A.V. (1965): The Peculiarity Index, a new function for use in Numerical Taxonomy. *Nature, Lond.*, *206* : 952.
3. Goodall, D.W. (1966): Deviant Index: a new tool for Numerical Taxonomy. *Nature, Lond.*, *210* : 216.
4. Marcus, L.F. and Vandermeer, J.H. (1966): Regional Trends in Geographic Variation. *Syst. Zool.*, *15* : 1-13.
5. Camin, J.H. and Sokal, R.R. (1965): A method for deducing branching sequences in Phylogeny. *Evolution*, *19* : 311-326.
6. Sokal, R.R. and Sneath, P.H.A. (1963): Principles of Numerical Taxonomy. San Francisco, W.H. Freeman.
7. Hall, A.V. (1967): Methods for demonstrating Resemblance in Taxonomy and Ecology. *Nature, Lond.*, *214* : 830-831.
8. Hall, A.V. (1967): Studies in recently developed group-forming procedures in Taxonomy and Ecology. *Journ. S. Afr. Bot.*, *33* : 185-196.
9. Cain, A.J. and Harrison, G.A. (1958): An anlysis of the taxonomist's judgement of affinity. *Proc. Zool. Soc. Lond.*, *131*: 85-98.
10. Sokal, R.R. and Michener, C.D. (1958): A statistical method of evaluating systematic relationships. *Univ. Kansas Sci. Bull.* *38* : 1409-1438.
11. Colless, D.H. (1967): An examination of certain concepts in Phenetic Taxonomy. *Syst. Zool.* *16* : 6-27.
12. Fisher, R.A. (1936): The use of multiple measurements in Taxonomic problems. *Ann. Eugen.*, *7* : 179-188.
13. Hall, A.V. (1968): Methods for showing Distinctness and aiding Identification of Critical Groups in Taxonomy and Ecology. *Nature, Lond.*, *218* : 203-204.

Discussion

Q. How was the abundance of species scored? By listing everything in the quadrangle?

A. This is a great problem, judging abundances. There are various techniques for it. The amount of cover can be one; the amount of basal cover, the amount of crown cover, the number of individuals, all these can provide fairly general estimates. We have to work within each particular framework.

Q. (Parker-Rhodes) Are you right in giving zero homogeneity to absent species? Surely it should be a mean value.

A. Yes. This is a great difficulty, actually. For species which are not at a particular site, absence may be fortuitous, the result of particular ecological situations, or even too small a sample area.

Comment: (Sneath) This is a problem for which some noughts are noughtier (naughtier) than others. (Great laughter).

THE PHENETIC RELATIONSHIPS BETWEEN SPECIES OF ONONIS

R. B. Ivimey-Cook

Department of Botany, University of Exeter

There are obvious advantages and disadvantages in presenting the opening contribution to this Numerical Taxonomy Colloquium. While I have to hope that what I shall be saying shall be along the lines that you wish to hear, at least I shall be able to make the more platitudinous remarks without fear that they have been contributed by the previous speakers.

Most of the reasons behind the choice of *Ononis* as the basis of the investigations are not relevant, except for one; many of the exercises in Numerical Taxonomy are made with a view to clarifying a difficult taxonomic position - quite rightly so if one had confidence in the methods adopted for the clarification. In the case of *Ononis*, there is in existence a monograph, written by Sirjaev in 1932, which, to say the least, is probably no worse than many other monographs. These investigations were made, therefore, not with a view to amending the infrageneric limits of *Ononis*, but to compare the results of a Numerical Taxonomic study with the conclusions of a taxonomist.

First, let me say a few general words about the genus, a conspectus of which is included as an appendix to this contribution. The Tribe Trifolieae in the Papilionaceae contains 6 genera from which *Ononis* may be distinguished by its relatively large corolla, which may be pink, purple or yellow in colour, and a legume which is neither coiled nor crescentic. Sirjaev had divided the genus into two Sections, *Natrix* to include those species in which the inflorescence is paniculate (each axillary peduncle bears 1 - 3 shortly pedicellate flowers) and the legume is elongated and pendant; *Bugranae* (properly *Ononis*) where the inflorescence is racemose and the legume ovoid or rhomboid and erect. Within both sections, a considerable number of sub-sections and series are recognised. The genus contains about 70 species (68 were actually used), mostly distributed along both shores of the Mediterranean. The centre of distribution seems to have been southern Spain or North Africa, where nearly half the species occur. Very few penetrate north of the Alps, though 3 species occur in Britain. Several others can be found in south-west Asia and the Middle East and one penetrates as far east as India. Generally, the species flourish in fairly dry and frequently calcareous soils, but they are not especially adapted to extremes of either temperature or humidity.

There is a considerable range in the species concept within the genus. Three species in particular are very variable - *O. natrix*, *O. viscosa* and *O. spinosa*. In the last, some of the variants are commonly treated as having specific rank, but in general, the variation is recognised by the establishment of a number of sub-species and varieties. At the other extreme, *O. aragonensis* and *O. reuteri* are distinguished by the size of their leaves and flowers.

Of the 68 species considered, material of 1 was not available in sufficient quantity for all the necessary determinations to be made. The 67 remaining were divided into 57, each of which was represented by 1 O.T.U. and 10 of the most variable were represented by more than one, in general by considering the next lowest acceptable taxon. Thus in the case of *O. natrix*, this was represented by 11 O.T.U.'s, the 11 sub-species.

For each O.T.U. 64 characters were investigated and their states determined. These 64 characters covered most of the morphological features of the species, but did not cover anatomical or physiological aspects - they thus took no account of information additional to that which Sirjaev would have had available. The determination of the states of the various characters was carried out on herbarium material chiefly in the British Museum and at Kew. For most of the quantitative characters there is a considerable range, not only on a single individual but also between individuals of one O.T.U. Since one has to decide on a single value for each cell of a data matrix, the mean (mode) of the values measured was adopted as the most satisfactory entry.

For certain characters, especially lengths and breadths, small differences at the low end of the range may be as important or significant as much larger differences at the upper end. Kendrick (1964) in his consideration of this problem, suggested resolving it by coding such states on a linear scale and incrementing the code increments. This approximates to a logarithmic transformation of the data. In the present case, logarithmic transformations were applied to the states of a number of the characters where it seemed that in the raw data such transformations would be appropriate. It is difficult to produce an absolute justification for this course of action in every case; on the other hand, there is no special virtue in the linear scale of measurement conventionally used.

Two main approaches may be used to obtain a classification of individuals - the divisive-monothetic and the agglomerative-polythetic. The former considers the entire population and splits it up hierarchically on the basis of the presence/absence of one attribute at every step of the hierarchy; the latter employs indices of similarity to determine which two individuals are most similar and these are combined, again hierarchically, until the original population has been reconstructed.

One divisive-monothetic method was investigated - the process of Association Analysis which has been used with considerable success in ecological investigations. The method utilises one of several possible association indices which have been constructed for use with presence/absence data; thus to apply the method it is necessary to recode the character states so that they conform to this requirement. Such a procedure is unsatisfactory, especially with quantitative data, despite the various methods which have been suggested whereby it can be carried out. As used, the method is incapable of splitting up a group of fewer than 8 individuals, so it is to be expected that the ultimate groups would show varying degrees of heterogeneity and the greatest interest will centre on the attributes which have been chosen to make the divisions. Thus the hierarchy can be looked upon as representing a dichotomous key in a rather unusual form and if the process were to be continued so that each of the final groups represented only a single individual, then one should expect the characters chosen as division attributes to be the best characters for the construction of a key for identification purposes. I have in fact glossed over two aspects - the choice of association index and whether positive and negative associations between the characters (attributes) should be given equal weight. In a taxonomic, as distinct from an ecological, situation both of these require to be considered in conjunction with the method adopted for coding the character states.

Agglomerative-polythetic methods are considerably more flexible - especially in their allowing multistate characters to be used (the notable exception being Information Analysis). Such multistate characters may be used to calculate a variety of similarity coefficients, of which I shall consider distance and the product-moment correlation coefficient.

Taxonomic distance is calculated as the distance between a pair of O.T.U.'s in a Euclidean hyperspace generated by treating the character states as a set of Cartesian co-ordinates. Since the states are measured in a variety of units, some restriction on the lengths of the character axes is required. In the present case, they were standardised.

Various techniques are available for the investigation of matrices of both distances and correlations. Both can be assessed by Cluster Analysis; when the matrix is of relationships between O.T.U.'s this gives a rapid resumé, the result of which can be usually portrayed as a two-dimensional dendrogram. The reduction on dimensions is indicative of the extent to which information on other relationships is missing.

Cluster Analysis may be either weighted or unweighted - in the former the similarity between a cluster and the remaining individuals is not dependent on the number of O.T.U.'s comprising each individual whereas in the latter, the number of individuals is taken into account. The terms are not, perhaps, ideal, since it is not clear whether they apply to the individuals or to the stems of the dendrogram. Pair-group methods would appear preferable to Variable-group, since Variable group methods required an arbitrary criterion of the limit of a group; many groups are of two individuals anyway while the generation of large variable groups would not appear generally useful.

Quite apart from any weighting which may be involved, the calculations of distances between a cluster and the remaining individuals may be carried out in a number of ways. One of the most frequently used is by average linkage - the arithmetic average of the distances between the components of the cluster and the individual. However, as Lance & Williams (1966) have recently shown, this is but a special case of a general system in which the nature of the sorting strategy is determined by the values of 4 parameters. They argue that the form of the cluster analysis now becomes a function of the clustering process and ceases to be an inherent property of the data.

Although the assessment of a correlation matrix between O.T.U.'s can be carried out by Cluster Analysis, the calculation of such a matrix is open to criticism. Since it is a Q-type analysis the elements of the data matrix vectors are heterogeneous, even if the effect of the units of the character states are removed by standardising the rows before calculating the correlations between the columns (representing the O.T.U.'s).

However, the calculation of correlations between the rows (characters) of the data matrix is unexceptionable, but this does not, of itself, give a direct measure of the relationships between the O.T.U.'s.

To digress somewhat for a moment. Although Numerical Taxonomy requires that a large number of characters should be used to obtain an assessment of phenetic relationships, it is also requisite that the results should be displayed over relatively few dimensions. Cluster analysis is one such method; another is Factor Analysis, the particular aspect of which is very apposite to taxonomic data being Principal Component Analysis. Principal Component analysis investigates the relationships between a multi-dimensional array of variables, some of which may be correlated and hence at least partially redundant. The points representing the O.T.U.'s can be treated as forming a roughly hyperellipsoidal swarm and the method will extract a set of orthogonal, and hence uncorrelated, components, being the principal axes of this hyper-ellipsoid. The components are extracted in descending order of magnitude and if much of the variance is extracted by the first few components, the effect is to reduce the dimensions of the data. The initial relationship between the variables is defined by the standardised data matrix, where one can envisage the O.T.U.'s as points defined on a series of axes, the characters, by the values of the states. The actual calculation of the components is carried out on a matrix of correlations between the characters.

The components extracted by Principal Component analysis have associated with them a vector which gives the relation between that component (axis) and the original character axes. These have been termed the component loadings. If this vector of loadings is used to postmultiply the standardised data matrix the result is a matrix which gives the relation between each O.T.U. and the component axes; these have been termed the component specifications.

Principal Component analysis works most efficiently where the swarm of points is approximately multivariate normal. In certain cases, especially where the points represent O.T.U.'s, there may well be a tendency for two or more swarms to appear - *e.g.* a dumb-bell shaped distribution which might correspond to two sub-generic taxa. Such a distribution is most likely to appear along the first component, which effectively finds the long axis of the dumb-bell. Under such circumstances, the second and subsequent components are likely to be of doubtful taxonomic meaning, since they will be constructed in the hyperspace common to the swarms and will reflect some characters of both. Such a difficulty is not obvious in the matrix of component loadings from the analysis of the between-characters correlation matrix, only making its appearance when the distribution of the O.T.U.'s is considered.

Principal Component analysis was first applied to the between-characters correlation matrix obtained from the complete 107 x 64 data matrix and the specifications of the O.T.U.'s were calculated. The positions of the O.T.U.'s were plotted using the first two components as axes - the result is shown in Fig.1. It will be seen that there is quite a well-marked split of the O.T.U.'s into two groups which corresponds with the two Sections of the genus, with 4 O.T.U.'s appearing intermediate. Of these, *O. tridentata* (2), *O. fruticosa* (3) and *O. pubescens* (38) to a lesser extent are classified more appropriately on component 3; *O. cintrana* (49) is in any event, in a rather intermediate position and the only O.T.U. which appears misclassified is *O. minutissima* (56). However, it must be recognised that the first two components only account for rather less than a quarter of the variation and a perfect separation should not be expected.

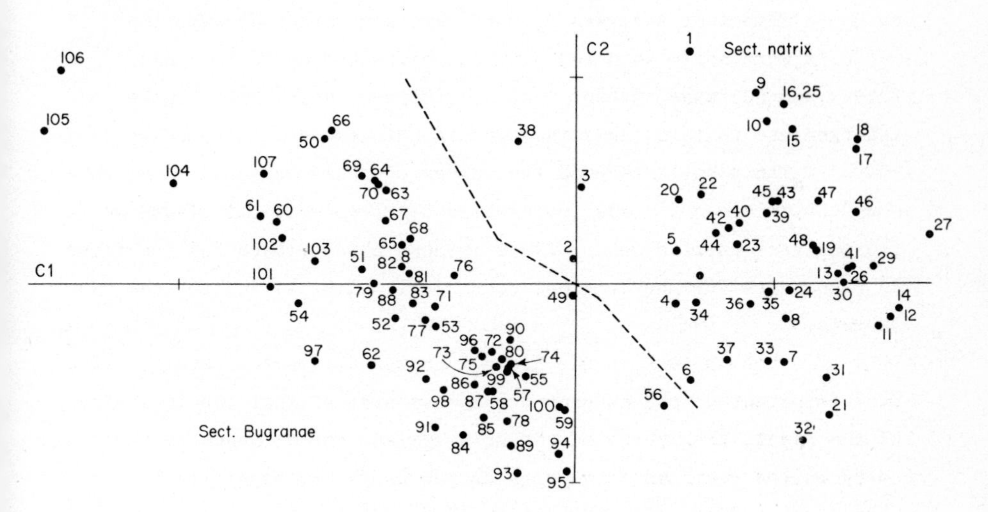

Fig. 1.

It is interesting to look also at the characters which are effective in bringing about this separation. These can be found from the component loadings, where characters with high loadings regardless of sign (i.e. with the component axis most nearly coincident with the character axis) are the most significant. The sign of the first element of the loading is apparently undefined (that is, the direction of the axis), but within the vector, the relative signs are always consistent. The important characters are the length of the peduncle and whether the pod is erect or pendant; these are both characters used to differentiate the Sections; to a lesser extent are the lengths of the arista and calyx tube, the number of nerves on the calyx tube and the number of seeds in the legume. The second component reflects chiefly the size of the foliage.

For the reasons that I have just mentioned, it was felt that the immediate recognition of these two sections might obscure the meaning of subsequent components and since it was not necessary to use a Component Analysis to make this separation anyway, the O.T.U.'s were split into two groups, corresponding to Sections Natrix and Bugranae, except that *O. cintrana*, which has a number of intermediate features, was included in both groups.

This split reduced the number of characters to 61 in group *N* and 58 in group *B*; new correlation matrices were calculated and the analyses carried out. Table 1 shows the variance for the first 12 components for the original genus (all O.T.U.'s) and for the two subsets.

Let us look at the Group *N* in rather greater detail. The most important of the component loadings suggest that the longevity of the plant, the length of the stipules and the presence of short hairs on the stem and long hairs on the calyx are significant characters so far as the direction of the first component axis is concerned, while for the second, the length of the arista, the yellowness of, and the absence of glandular hairs on, the standard petal. If the component specifications are calculated, the scatter diagram over the first two components is as shown in Fig. 2.

In this, two of the most prominent groups of O.T.U.'s have been indicated, such as *O. natrix* (O.T.U.'s 9 - 19) and *O. viscosa* (O.T.U.'s 39 - 45). Certain groupings of related species are also evidence, such as the *O. reclinata* group (O.T.U.'s 34 - 37), while several of the more distinctive O.T.U.'s are widely separated (49, which is the intermediate, *O. cintrana*, and also the 3 species in Sub-section Rhodanthae, with *O. rotundifolia* (O.T.U.'s 2 - 4 and 1).

Species Ononis

Component	COMPLETE		GROUP *N*		GROUP *B*	
	Variance	% of total	Variance	% of total	Variance	% of total
1	10.256	16.02	6.976	11.44	7.825	13.49
2	4.748	23.44	6.191	21.59	6.190	24.16
3	4.243	30.07	5.642	30.84	5.353	33.39
4	3.749	35.93	3.992	37.38	4.485	41.13
5	3.127	40.82	3.899	43.77	3.338	46.88
6	2.970	45.46	3.580	49.64	3.111	52.24
7	2.316	49.08	3.200	54.89	2.780	57.04
8	2.221	52.55	2.726	59.36	2.426	61.22
9	2.193	55.97	2.312	63.15	2.356	65.28
10	2.078	59.22	2.013	66.45	1.873	68.51
11	2.000	62.35	1.694	69.23	1.785	71.59
12	1.725	65.04	1.561	71.79	1.554	74.27
Total	41.624		43.789		43.077	

Table 1.

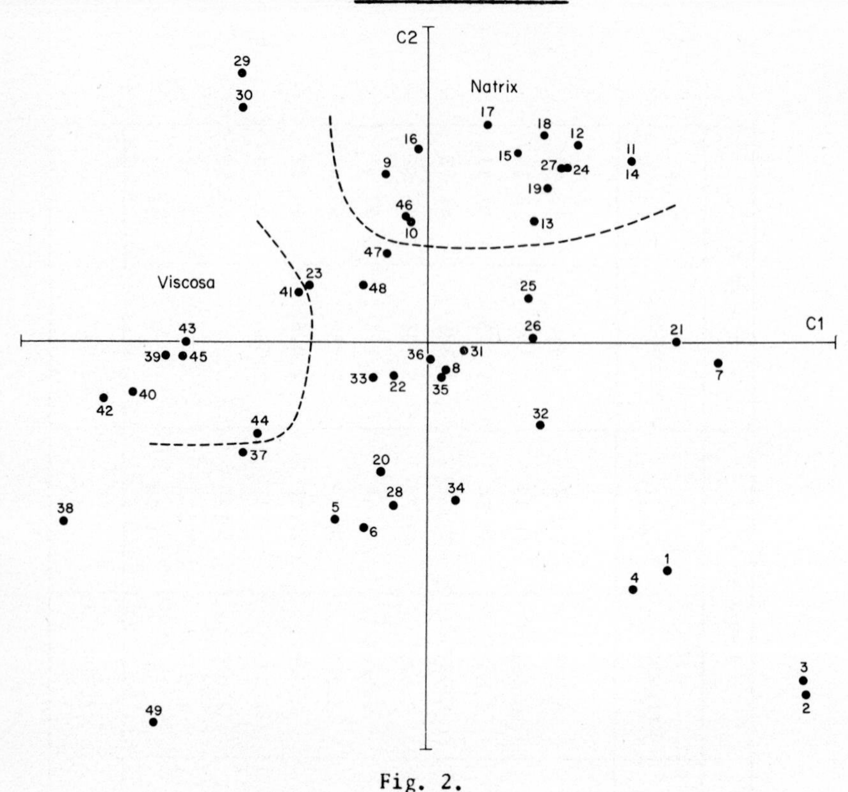

Fig. 2.

However, it is unrealistic to expect a good discrimination when only 21% of the variation has been accounted for.

A more adequate investigation of this distribution can be obtained if one simulates a model over, say, 12 dimensions - the first 12 component axes which account for 72% of the variation. Such a model cannot, of course, be visualised, but can be used to obtain a distance matrix between the O.T.U.'s which can then be subjected to Cluster Analysis; in general, the unweighted pair group method seems to give the best results in such cases. The dendrogram which can be constructed from these results is shown in Fig. 3. Examination of this shows quite clearly the groups of O.T.U.'s which have already been mentioned; particular points of interest is the marked distinction between 6 of the sub-species of *O. natrix* (*natrix*, *ramosissima*, *stenophylla*, *angustissima*, *polyclada* and *filifolia*) on the one hand and the remaining group comprising

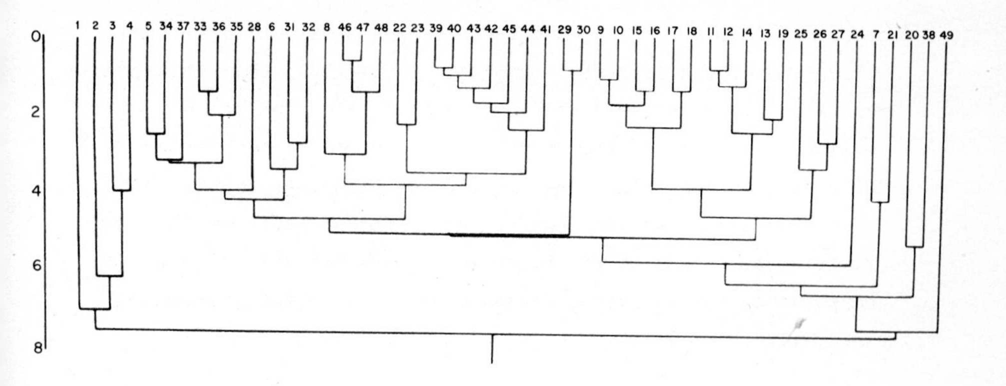

Fig. 3.

sspp. *hispanica, mauritii, arganietorum, prostrata* and *falcata* on the other. These two groups are quite clearly distinguishable in the field on the basis of the size of their leaves and stipules; another point is that the recent separation of *O. crotolarioides* (45) as a distinct species, chiefly on the basis of its much inflated pod, would not appear to be supported by these results. Taken overall, it would appear, to be no more dissimilar than the other sub-species of *O. viscosa*, of which it was once one. The higher levels of clustering are quite reasonably interpreted, though it is not possible to assign a distance at which it can be said that below this one is dealing with species and above it with a higher taxon. Since these larger aggregates are based on overall similarity, it is not to be expected that they will coincide particularly well with the Sub-sections and Series of Sirjaev, which are based on only a few characters.

An analogous series of operations were carried out on the other group (b). Again, one can show a scatter diagram of the O.T.U.'s over the first two components, (Fig. 4), the characters most strongly associated with component 1 being the breadth of the leaflet, the shape of the calyx and the length of the standard petal. O.T.U.'s with long standards, tubular or bilabiate calices and relatively large leaflets, which belong to Sub-sections Verae and Crinitae, are found at the negative end of this axis. For component 2, the presence of glandular hairs on the standard is especially prominent.

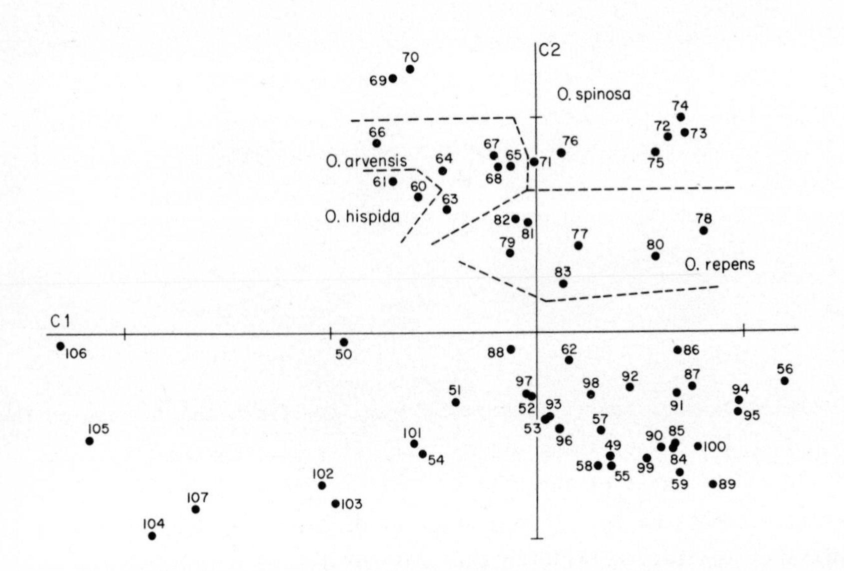

Fig. 4.

The dendrogram derived from a Cluster Analysis (by the unweighted method) from a distance matrix calculated over the first 12 component specification is shown in Fig. 5. One can notice especially that the infraspecific taxa tend to cluster together (*e.g.* those of *O. spinosa* (69 - 76); *O. repens* (77 - 82) and *O. arvensis* (63-68). In the case of *O. spinosa*, there appears to be no clear distinction between those O.T.U.'s which have been accorded varietal and those which have been accorded sub-specific status 72 - 74 are varieties of ssp. *antiquorum* while 76 is ssp. *foetens*); considering both *O. repens* and *O. arvensis* one might draw the conclusion that taxa treated as varieties are more different from each other than some of the sub-species of *O. spinosa*. The case for a thorough investigation here would appear unanswerable. It is also noteworthy that *O. masquillieri* which is generally classified with this group but has certain very distinctive features, completely fails to cluster with it. A number of other groups are clearly indicated, *e.g.* Sub-section Chrysanthae (55 - 59) and Series Tuberculatae in Sub-section Diffusae (91 - 95). The evaluation of the larger groups indicated by the dendrogram is rendered difficult by the marked chaining tendency. It suggests that there is probably a group of rather similar species represented by those in the upper third of the dendrogram, while the remainder form rather dissimilar groups, which must, perhaps unfortunately, be brought into the system one at a time.

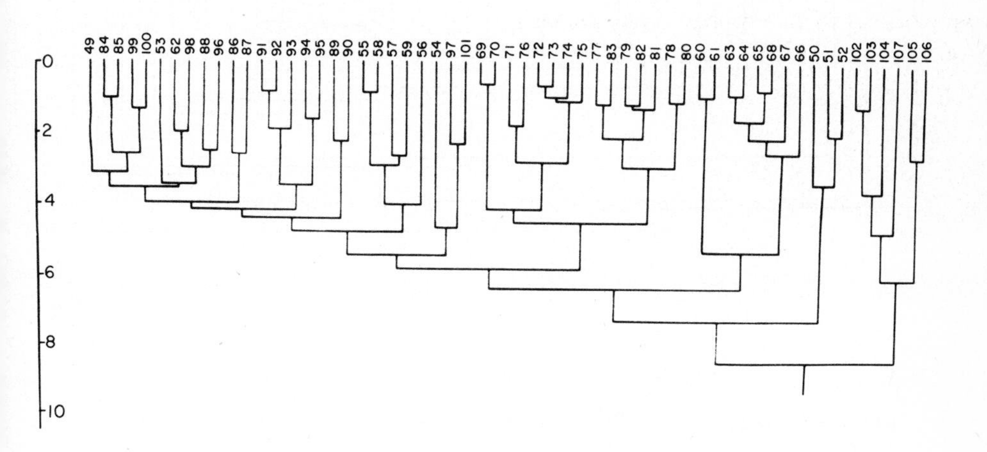

Fig. 5.

Distance matrices can, of course, be calculated directly between the individuals (O.T.U.'s) of a data matrix, the constraint on the heterogeneity of the O.T.U. vectors usually being standardisation. Such distances matrices are relatively easy to calculate and they can be processed by Cluster Analysis to give a dendrogram in which the apparent affinities of the O.T.U.'s is directly related to all the data collected into the data matrix. The cluster analysis is necessarily taking place over a much larger number of dimensions than in the case of the component specifications and the results often seem to be much more "messy", with a fairly substantial

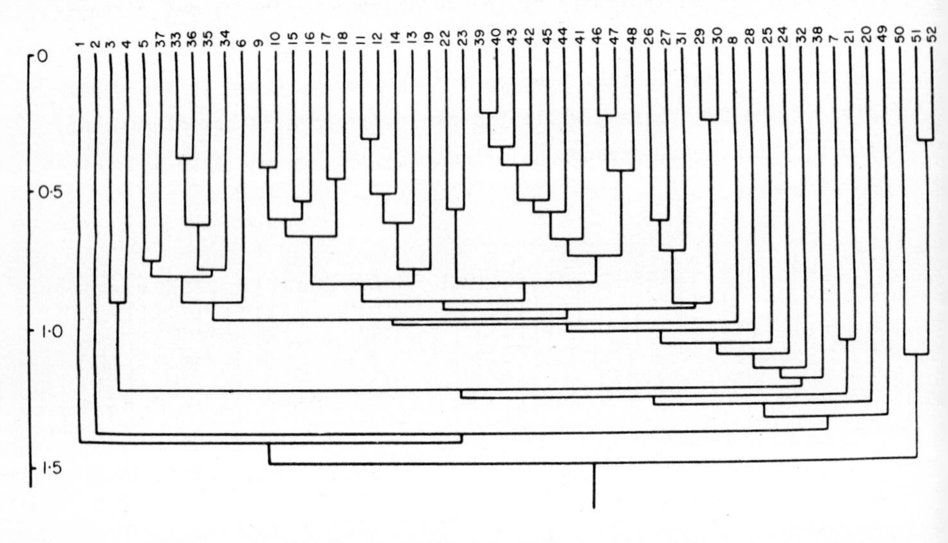

Fig. 6.

group of residual, or "joining by necessity" species. In the case of group *N* (Fig. 6), this is only partially true, the component specification dendrogram and the distance matrix dendrogram being almost identical; it is much more true in the case of the more complex group *B*(Fig. 7). One has the further disadvantage that one is deprived of any information, such as that provided by the component loadings, to indicate the basis for the results obtained.

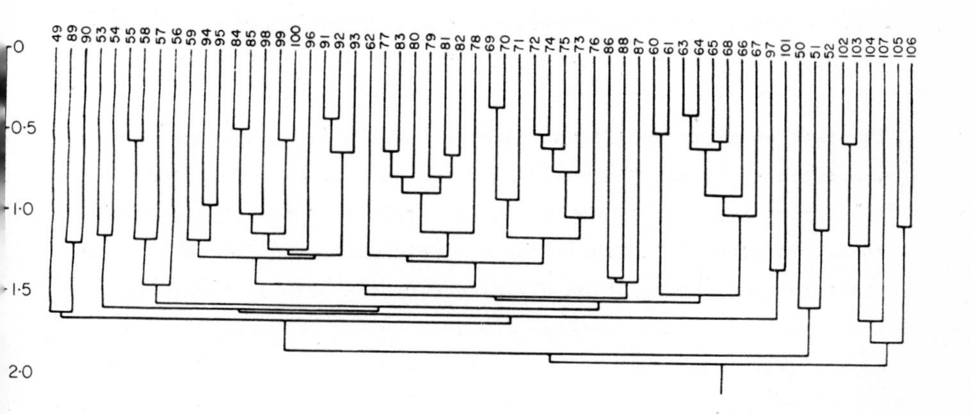

Fig. 7.

Discussion

It has been suggested that the introduction of numerical methods into taxonomy will remove the subjective element that has tended to obscure certain facets of the process of classification. This is only partially true and the result has been to divert this subjectivity, though reducing it somewhat in the process, from the assessment of similarity to the choice of the characters from which the assessment is made.

The 64 characters used in this investigation are probably a minimum, though much depends on the type of character. There seems no obvious justification for the arbitrary weighting of floral over vegetative or other characters, though some measure of such weighting will be introduced by the proportions of the various characters used.

Of the analytical methods used, Cluster Analysis is the simplest agglomerative polythetic method and the construction of 2-dimensional dendrograms presents the information in an easily assimilable form, but the reduction in the number of dimensions involves the loss of a considerable amount of information - all the information about relationships other than those expressly indicated. Cluster Analysis requires that at each stage of the clustering process, a new matrix is calculated, treating clusters as new individuals. The shape of the clusters so formed is important in assessing the efficiency of the method. Pair-group clustering by average linkage can lead to the centroid of a cluster moving away from other O.T.U.'s which are similar to one of the individuals in the cluster but not sufficiently similar to join it. Such O.T.U.'s are liable to become completely isolated and may eventually join a cluster to which they are not especially closely related. This type of tendency is particularly evident in Fig. 6. While Variable-group methods help to overcome this problem, the arbitrary limit of a group is open to objection; it is also not able to deal effectively with clusters which are not approximately hyperspherical. In taxonomic data, the possibility of elongated clusters is a real one, though it is difficult to devise a method which will locate these accurately. A recent paper by Carmichael et al (1968) suggests a method of clustering which may resolve this difficulty.

One of the problems which arises when the results of different methods are to be compared is the mechanism of the comparison. Sokal and Sneath (1963) suggested the use of the cophenetic correlation coefficient; while this will always recognise two identical dendrograms or other patterns, the amount of dissimilarity permitted before the dendrograms cease to be identical must depend on the precision of the comparison. With taxonomic data, a certain and perhaps considerable variation, must be expected, but there appears to be no statistical test which will take account of this.

A general comparison of these results with the system proposed by Sirjaev suggests that there are few significant discrepancies, though a case could perhaps be made for the erection of two new Sections, one to contain *O. rotundifolia* and the other to contain the Sub-sections Salzmannianae, Verae and Crinitae. Both Sections could be readily identified on morphological grounds.

References

Carmichael, J.W., George, J.A., Julius, R.S. (1968). Finding Natural Clusters. *Syst. Zoo. 17*, 144-150.

Kendrick, W.B. (1964). Quantitative characters in computer taxonomy. In Phenetic and Phylogenetic Classification Publ. *Syst., Assn. 6*, 105-114.

Lance, G.N. & Williams, W.T. (1966). A generalized sorting strategy for computer classifications. *Nature, 212*, 218.

Sirjaev, G. (1932). Generis Ononis revisio critica. *Beih. Bot. Centr. 49*, 381-665.

Sokal, R.R. & Sneath, P.H.A. (1963). Principles of Numerical Taxonomy. Freeman.

Conspectus of the Genus Ononis (after Sirjaev, 1932)

O.T.U. Number

NATRIX

Raceme pedunculate, 1-3 flowered; fruiting pedicel reflexed; pod linear or oblong.

Antiquae

Subshrub, leaves and bracts 3-foliolate; vexillum hairy; 10th stamen adnate at the base only . . . . *rotundifolia* 1

Rhodanthae

Shrub or perennial; leaves ternate, coriaceous; pedicel shortly aristate; vexillum hairy.

Frutescentes

Shrub; bracts bract-like; pedicel short, 1-3 flowered . . . . *tridentata* 2
fruticosa 3

Perennes

Perennial; bracts leafy; pedicel mucronate, 1-flowered, occasionally with 2 . . . *cristata* (cenisia) 4

Canarienses

Shrub; leaves 3-foliolate; peduncle short, muticous; corolla rose, glabrous; alae without teeth; seeds tuberculate . . . *christii* 5

Mauritanicae

Shrub; leaves pinnate, bracts 3-foliolate; corolla rose, glabrous; alae without teeth . . . . *thomsonii* 6

Eu-natrix

Shrub or perennial; leaves usually 3-foliolate, but variable; bracts variable; corolla yellow, sometimes red-striped, glabrous; alae without teeth.

Atlanticae

Peduncle spinous-aristate; stipules not connate . . . . *atlantica* 7

Orientales

10th stamen one-third adnate; lower leaves pinnate; peduncle usually 2-flowered . *adenotricha* 8

Polymorphae

Peduncle unarmed; stipules not connate; 10th stamen two-thirds adnate . . . . *natrix* 9-19
crispa 20

Aegypticae

Shrub; stipules connate; peduncle subspiny; leaves 3-foliolate . . . *vaginalis* 21

Serotinae

Perennial; leaves 1-foliolate; peduncle 1-flowered, muticous or shortly aristate . . . . *serotina* 22
pseudoserotina 23

Torulosae

Annual; corolla yellow, glabrous; pod torulose; seeds tuberculate . . . *ornithopodioides* 24

Biflorae

Annual; peduncle 2-flowered, aristate; corolla yellow, glabrous.

Tuberculatae

Seeds tuberculate . . . . *biflora* 25

Laeves

Seeds smooth . . . . *hebecarpa* 26

Pilosae

Leaves 3-1-foliolate; corolla rose and yellow, vexillum hairy; alae without teeth . . . . *polysperma* 27
maweana 28

Reclinatae

Annual; peduncle 1-flowered; vexillum glabrous; alae without teeth.

Siculae

Corolla yellow; peduncle aristate; seeds tuberculate . . . . *sicula* 29-30

Sublaeves

Corolla yellow; peduncle aristate; seeds nearly smooth . . . . *peyerimhoffii* 31

Incisae

Corolla yellow; peduncle shortly aristate; seeds rugose; stipules incised . . . . *incisa* 32

Eu-reclinatae

Corolla rose or whitish; peduncle muticous; seeds tuberculate . . . . . *reclinata* 33
pendula 34
laxiflora 35
dentata 36
verae 37

Viscosae

Annual; leaves mostly 3-foliolate; corolla yellow, glabrous; alae without teeth.

Pubescentes

Upper bracts bract-like; peduncle muticous; seeds smooth . . . . *pubescens* 38

Eu-viscosae

Bracts leafy; peduncle aristate; seeds tuberculate . . . . *viscosa* 39-44
crotolarioides 45

Antennatae

Leaves all 3-foliolate; peduncle aristate; seeds tuberculate . . . . *antennata* 46-48

BUGRANAE

Peduncle nearly or quite absent; pod ovate, rarely reflexed.

Intermediae

Annual; vexillum hairy; pod reflexed; seeds tuberculate . . . . *cintrana* 49

Chrysanthae

Shrubby; leaves 3-foliolate; bracts bract-like; flowers yellow; seeds smooth.

Speciosae

Leaflets subcoriaceous; raceme dense . *speciosa* 50

Aragonenses

Leaflets coriaceous; raceme interrupted . . . . *aragonensis* 51

Pinnatae

Annual, shrubby; leaves pinnate; corolla rose, glabrous; seeds smooth . . . . . *leucotricha* 53
pinnata 54

Section / description	Species	Page
Bugranoides		
Shrubby or perennial; leaves 3-foliolate corolla yellow, glabrous	*pusilla*	55
	(columnae)	
	minutissima	56
	cephalotes	57
	saxicola	58
	striata	59
Acanthononis		
Shrubby or perennial, often spiny; leaves variable; corolla rose; hairy; seeds rarely smooth.		
Arborescentes		
Small unarmed shrub	*hispida*	60-61
Vulgares		
Shrubby, often spiny	*masquillieri*	62
	arvensis	63-68
	(hircina)	
	spinosa	69-76
	repens	77-83
Villosissimae		
Annual; leaves 1-3-foliolate; corolla rose, glandular-hairy; seeds tuberculate.		
Trifoliatae		
Leaves all 3-foliolate	*filicaulis*	84
	villosissima	85
Monophyllae		
Leaves all 1-foliolate	*monophylla*	86
	oligophylla	87
	alba	88
Variegatae		
Annual; leaves 1-foliolate; flowers in lax racemes; corolla yellow, hairy	*variegata*	89
	euphrasiaefolia	90
Diffusae		
Annual; leaves 3-foliolate; corolla rose or white, usually hairy; raceme elongating after anthesis.		
Tuberculatae		
Vexillum hairy; seeds tuberculate . .	*diffusa*	91-93
	serrata	94
	phyllocephala	
	tournefortii	95

Section / characters	Species	No.
Cossonianae		
Vexillum glabrous; seeds smooth . . .	*hirta*	96
	cossoniana	97
	cephalantha	98
Subspicatae		
Vexillum glabrous; seeds tuberculate .	*subspicata*	100
Mitissimae		
Annual; leaves 3-foliolate; upper bracts imbricate, coriaceous; corolla rose, glabrous; calyx tubular; seeds tuberculate	*mitissima*	101
Salzmannianae		
Annual; leaves 3-1-foliolate; calyx tubular; corolla rose, glabrous; seeds smooth	*alopecuroides*	102
	baetica	103
Verae		
Annual; leaves 3-foliolate; upper bracts bract-like; calyx tubular; corolla large, rose; vexillum hairy; seeds tuberculate	*rosea*	104
	avellana	105
	megalostachys	106
Crinitae		
Annual; leaves and bracts 3-foliolate; calyx bilabiate, tubular; corolla whitish, glabrous; seeds tuberculate	*crinita*	107

COMPARISON OF CLASSIFICATIONS

D. M. Jackson

Mathematical Laboratory, University of Cambridge

1. Introduction

The work to be described here arose from a set of experiments in information retrieval in which the effect on performance of different classifications and different descriptions of the same document collection was being examined (Spark Jones, Jackson[1]). The classifications used were generated automatically and consisted of non-hierarchical overlapping sets of keywords. The problem arose of finding a way to describe how a pair of classifications differ. In addition it was necessary to attempt to establish a 'relationship' between a classification and the information classified and to express the result in a form which would provide an estimate of the reliability of that relationship.

2. Conditions on classification and property arrays

In the particular application described above, the classes consisted of keywords and were generated using estimated resemblances or similarities between keywords evaluated from their co-occurrence in the document descriptions. In this discussion it is more convenient to forget the particular application and to use more general terminology. Thus classes are to consist of objects and these objects are described in terms of properties. The properties are binary in that their presence or absence only is recorded. Similarly, class membership is a two valued relation in that an object either belongs or does not belong to a class. Neither class membership nor the possession of a property may be quantified by probabilistic weights. Both the original information and the classification may therefore be represented completely by binary arrays. The rows of the arrays represent either classes or properties and the columns represent objects.

Suppose that D is the binary array for the data, and C is the binary array for the classification.

Then $d(i,j) = 1 \iff$ object j has property i

$= 0 \iff$ object j does not have property i

Also $c(i,j) = 1 \iff$ object j belongs to class i

$= 0 \iff$ object j does not belong to class i

Suppose that we now consider a specific object j and obtain its description both in the original data and the classification. We shall say that the class description of j is more economical than the property description of j provided that

$$N(i|d(i,j) \neq 0) > N(i|c(i,j) \neq 0) \qquad (1)$$

where the notation $N(x|\langle\text{condition on } x\rangle)$ is the number of entities x which satisfy <condition on x>.

The uses to which classifications are put are varied but they generally have in common the fact that they may be employed to make inferences about the population of objects from the classification of these objects. Suppose that we consider the objects i and j and note that they tend to occur together in the same classes. We would hope that the inference that objects i and j are closely related to each other would be confirmed by their property descriptions in the original data. This remark requires qualification for it may not hold in every case. However, if it does not hold in a significant proportion of the cases in which the class descriptions are comparable, then some further steps would be required to justify regarding the grouping of objects as a classification. A stronger remark, made with the same qualification, is to require that the relatedness of a pair of objects estimated from their class descriptions may be inferred from their relatedness as estimated from their property descriptions. Thus we require that

(i,j) strongly related in $D \iff (i,j)$ strongly related in C (2) .

An example of this is found in information retrieval. Suppose for example, that a group of descriptors tend to occur together in the same group of documents. Then for the purposes of matching documents against a particular request these descriptors are inter-substitutable. This group of descriptions would hopefully be isolated by classification. If they were isolated then they would be regarded as closely related (or inter-substitutable in retrieval applications). The retrieval performance would mirror the extent to which this hypothesis was supported in reality by the document descriptions.

Let us suppose that we have a function h which measures the relatedness or similarity between pairs of objects described by binary properties (Ball[2]). For each pair of objects two estimates are available, one from the data array, D, and the other from the classification array C. The distribution of similarities derived in this way will be different for the two arrays so that in particular a similarity which is strong relative to the distribution in D is not necessarily strong relative to the distribution in C. The direct comparison between coefficients derived from different arrays as implied by (2) must be avoided. Instead, we reformulate (2) in terms of a comparison between similarities derived from the same array. Thus we require that

$$\text{sign}\ (S(i,j)-S(k,\ell)) = \text{sign}\ (T(i,j)-T(k,\ell) \qquad (3).$$

where S and T are similarity matrices derived from D and C respectively by applying h. The proportion of times (3) holds to the number of distinct quadruplets $(i,j;\ k,\ell)$ taking due account of symmetries between the four integer variables is called the 'discrepancy'. It is maximum when (3) fails in all cases. Although this gives an overall impression of the confidence with which we make inferences between the two arrays of the type described above, a more useful guide is provided by determining the discrepancy first for the most strongly related pairs and then for the less strongly related pairs and so on to the least strongly related pairs. The simplest way of achieving this is to calculate the discrepancy for all pairs of objects which have a similarity above a particular threshold in the two matrices and to vary this threshold from the minimum to the maximum. The direct comparison between values of the similarity coefficients in the two matrices implied by applying a single threshold to them both is avoided by considering the proportions into which an arbitrary threshold divides the matrices. Accordingly, suppose that s is the threshold applied to S and that t which depends on s, is the threshold applied to T.

Let $P(S,s)=N((i,j)|S(i,j)>=s)$ and $Q(S,s)=N((i,j)|S(i,j)<s)$

and $P(T,t)=N((i,j)|T(i,j)>=t)$ and $Q(T,t)=N((i,j)|T(i,j)<t)$

Then choose $t(s)$ to minimise

$$|(P(S,s)/(P(S,s)+Q(S,s)) - (P(T,t)/(P(T,t)+Q(T,t)))| \quad (4)$$

3. The discrepancy and transition categories

The set of ordered pairs of objects is divided into two exclusive groups by (3). Those which satisfy the condition are called positive transitions and those which do not are called negative transitions. The proportion of positive transitions to the number of distinct pairs is called the discrepancy. We introduce the discriminant of a pair of objects with respect to threshold s in S and $t(s)$ in T defined by

$$u(i,j;k,\ell;s)=f((S(i,j)-S(k,\ell))/(T(i,j)-T(k,\ell))) \quad (5)$$

where

$$\begin{aligned} f(x) &= +1 \text{ if } x > 0 \\ &= -1 \text{ if } x < 0 \\ &= 0 \text{ if } x = 0, \text{ or } x \text{ is infinite,} \\ &\qquad \text{or } x \text{ is undefined.} \end{aligned}$$

It should be noted that $u(i,j;k,\ell;s)$ is symmetrical in (i,j) and (k,ℓ). The discriminant separates pairs of objects in greater detail than does (3). If we consider ordered pairs of objects there are nine categories generated by (5), corresponding to the numerator and denominator independently being positive, negative or zero. The categories are denoted by the sign of the numerator followed by the sign of the denominator, or zero in the case of equality.

Thus (+,+) and (-,-) correspond to positive transitions.
(-,+) and (+,-) correspond to negative transitions.
(0,-) and (0,+) correspond to ambiguous transitions.
(-,0) and (+,0) correspond to ambiguous transitions.
(0,0) corresponds to ambiguous transitions.

Since $u(i,j;k,\ell;s)$ is symmetrical in (i,j) and (k,ℓ) the distinction between (+,+) and (-,-) etc is not made although it could in principle be made if conditions were imposed on (i,j) and (k,ℓ) enumerate all distinct pairs $((i,j),\ (k,\ell))$ once only. This would provide evidence on the comparative size of the probabilities of $D = C$ and $C = D$. The five categories which remain after coalescing those which are indistinguishable by symmetry are denoted by $+T$, $-T$, EX, XE, EE respectively, where $+T$ and $-T$ are the positive and negative transitions, E stands for 0 and X stands for + or -.

The positive transitions are given by

$$R(+,s) = N((i,j;k,\ell)\ u(i,j;k,\ell,s) = +1;\ \text{condition}) \qquad (6)$$

The negative transitions are given by

$$R(-,s) = N((i,j;k,\ell)\ u(i,j;k,\ell;s) = -1;\ \text{condition})$$

The ambiguous categories are given by

$$R(0,s) = N((i,j;k,\ell)\ u(i,j;k,\ell;s) = 0;\ \text{condition})$$

where the condition is

$$\min(S(i,j),\ S(k,\ell)) >= s;\ \min(T(i,j),T(k,\ell)) = t(s);\ 1 < i,j,k,\ell < n.$$

The ambiguous transitions may either be regarded as positive or as negative transitions. Accordingly there are two values of the discrepancy g defined by

$$g(-,s)=(R(-,s) + R(0,s))/(R(-,s) + R(+,s) + R(0,s)) \qquad (7)^*$$

and $$g(+,s) =R(-,s)/(R(-,s) + R(+,s) + R(0,s))$$

corresponding to assigning ambiguous transitions to negative transitions and to positive transitions respectively.

* The same measure is proposed by M.G. Kendall,(3), and is called the Coefficient of Agreement. It is scaled to lie between -1 and +1. He is however, not concerned with the computational aspects of this measure for large preference tables and, more seriously, does not encounter the problem of ambiguities, which would occur in the circumstances he describes if there were individuals who refused to express preferences in certain cases.

Here is an example:- Suppose that S and T are two symmetrical matrices with zero leading diagonals defined as

$$S = \begin{vmatrix} 0 & 1 & 3 & 10 & 15 \\ 1 & 0 & 1 & 6 & 2 \\ 3 & 1 & 0 & 5 & 8 \\ 10 & 6 & 5 & 0 & 3 \\ 15 & 2 & 8 & 3 & 0 \end{vmatrix} \quad \text{and} \quad T = \begin{vmatrix} 0 & 3 & 5 & 6 & 7 \\ 3 & 0 & 6 & 5 & 1 \\ 5 & 6 & 0 & 7 & 9 \\ 6 & 5 & 7 & 0 & 8 \\ 7 & 1 & 9 & 8 & 0 \end{vmatrix}$$

Let the pair (i,j) where $i<j$ be denoted by the point Pk where $k=j-(i(i+1)/2) + n(i-1)$. The partial orderings represented by these matrices for pairs (i,j) where $i<j$ are

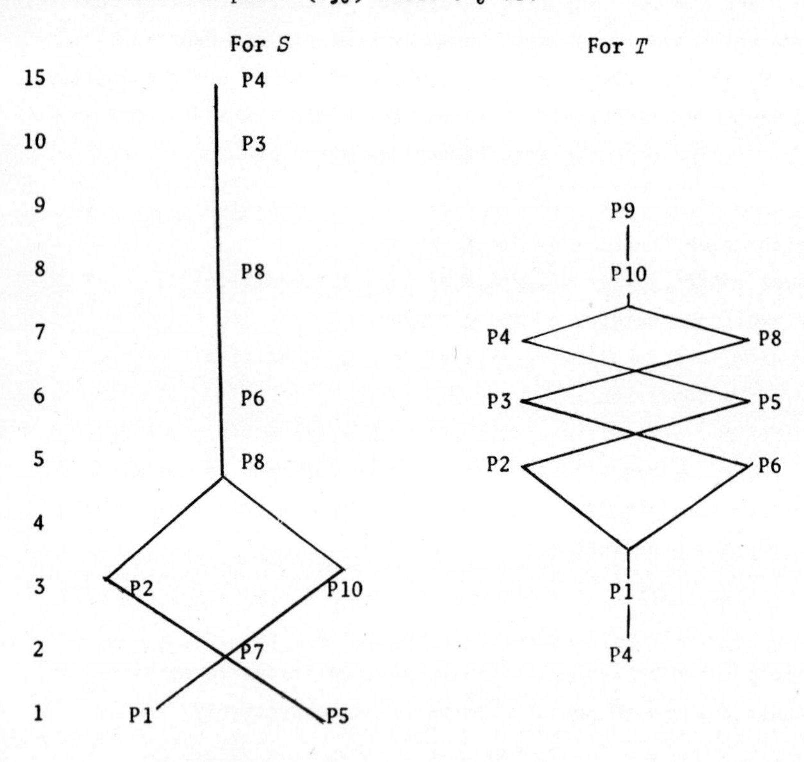

Fig. 1 Partial ordering for arrays D and C with hypothetical similarity function

The transitions enumerated below are calculated for s=0 and t=0.

	P	Points in T									
		1	2	3	4	5	6	7	8	9	
p	2	$+T$									
t	3	$+T$	$+T$								
s	4	$+T$	$+T$	$+T$							
	5	EX	$-T$	XE	$+T$						
i	6	$+T$	XE	$+T$	$+T$	$-T$					
n	7	$-T$	$+T$	$-T$	$+T$	$-T$	$+T$				
	8	$+T$	$+T$	$-T$	XE	$+T$	$-T$	$+T$			
S	9	$+T$	$+T$	$-T$	$-T$	$+T$	$+T$	$+T$	$+T$		
	10	$+T$	EX	$-T$	$-T$	$+T$	$-T$	$+T$	$-T$	$+T$	
T	$+T$	7	5	3	3	3	3	3	1	1	28
O	$-T$	1	1	3	2	2	1	0	1	0	12
T	EX	1	1	0	0	0	0	0	0	0	2
A	XE	0	1	1	1	0	0	0	0	0	3
L	EE	0	0	0	0	0	0	0	0	0	0

Fig. 2. Transition categories for the partial orderings above

The transition categories for (P_i, P_j) *wrt* matrices S and T are shown in the table together with the totals for each P_i in T having been compared with every P_j in S. The overall totals for each category are given in the last column of the totals. The number of entities in each category are as follows $(R+,0)$ = 28, $R(-,0)$ = 12, $R(0,0)$ = 5. Therefore the discrepancies are $g(-,0)$ = 0.73 and $g(+,0)$ = 0.62.

4. Computational procedures

The similarity matrices we are concerned with may be large compared with the core store capacity of the machine. In many cases, however, the proportion of non-zero elements to the total number of elements is small and advantage may be made of this when computing the discrepancy. Suppose that n is the dimension of the matrices. The density of a matrix is defined to be the proportion of non-zero elements to the total number of elements in the matrix.

Thus $$d = N((i,j)\,|\,S(i,j)>s)/N(i,j)) \qquad (8)$$

Here we are concerned with the density relative to a threshold s, that is the proportion of elements larger than s to the total number of elements. Recent work has been concerned with matrices S where $n = 712$ and $d(S,0) = 0.09$ and with $n = 351$ and $d(S,0) = 0.90$. Let Z be the approximate time taken to calculate the discrepancies for m thresholds. Then an approximation to Z is $Z = KmQe$ where Q is the total number of quadruplets $(i,j;k,\ell)$, K is the time taken to calculate a single transition and e is the mean instruction time. The practical constraints of a possible algorithm are

i) it must deal with matrices which are large compared with the available store.

ii) the order of the process may be at most 2 in n.

Now K is smallest when the matrices are laid out row by row in the store, when $S(i,j)$ and $T(i,j)$ may be located directly. However, n may be larger than the amount of store so this is not possible. Instead, thresholds s and $t(s)$ are chosen so that only the elements of the matrices which are larger than s and t are stored. K is larger in this case but may be reduced by ensuring that the column numbers for each row are sorted into numerical order of column numbers. The number of quadruplets $(i,j;k,\ell)$, ignoring symmetries, decreases and becomes

$$Q = n^2(max(d(S,s),\ d(T,t)).$$

This may be reduced if the symmetries between i and j, between k and ℓ, and between (i,j) and (k,ℓ) are taken into account. In this case

$$\begin{aligned} Q &= N((i,j;k,\ell)\,|\,i>j;\ k>\ell;\ 1=j=>k>i;1\nmid j;\ell=<i,j,k,\ell=<n) \\ &= (n+1)n(n-1)(n-2)/8 \end{aligned}$$

For sufficiently large n, this gives a reduction by a factor of eight. However, Z is of order 4 in n. Now suppose that we deal directly with pairs (i,j) rather than with row index i and column index j separately. The matrices are stored row by row and all trace of the original row-column structure is removed. At the same time the matrices are intersected. Thus both $S(i,j)$ and $T(k,\ell)$ are removed if $S(i,j)<s$ or $T(i,j)<t$. Suppose that as a result there are n' elements in each of the intersections S' and T'. Then $n' < n\ min(d(S,s),d(T,t))$. By forming the intersection of S and T the address of element (i,j) in T' may be derived quite simply from the address of (i,j) in S' as $((i,j)$-base$(S))$ + base(T) where base(S) and base(T) are the start addresses of S and T. Thus we may behave as if the matrices were laid out in the core store in their complete form after thresholding. K is therefore, minimal (say three machine orders), and the number of pairs $((i,j),(k,\ell))$ is now $n'(n'+1)/2$. The process is now of order 2 in n. In addition, the discrepancies for all of the n thresholds may be computed simultaneously. This is done by keeping separate counters for each transition and for each threshold combination $(s,t(s))$. For each discriminant the counters corresponding to

$$min(S(i,j),S(k,\ell),T(i,j),T(k,\ell))$$

are updated. When all the discriminants have been calculated the actual numbers of transitions for each category for each threshold s' are computed by summing the number of transitions in each category for every threshold larger than s'. Such an algorithm takes time $Z=KmQe$ where

i) $m = 1$

ii) $K = 3$

iii) $Q = n'(n'+1)/2$ where $n'<nn.\max(d(S,s),d(T,t(s))$

Thus the algorithm is usable provided that $d(S,s)$ and $d(T,t(s))$ are sufficiently small. This condition in certain cases may require that the initial choice of s and t causes the removal of a large proportion of the possible thresholds above each of which the discrepancies are to be calculated. The following procedure is not subject to this restriction.

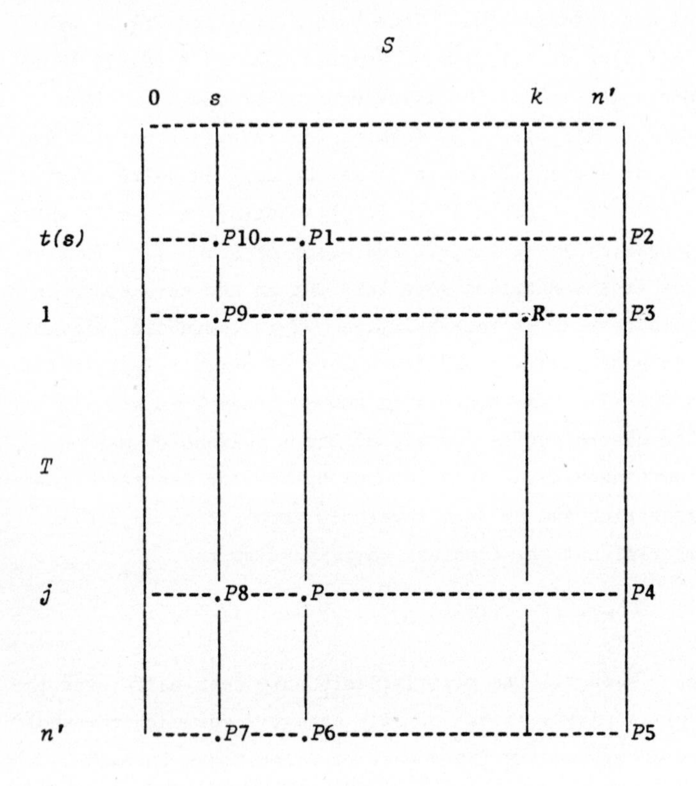

Fig. 3 Incidence matrix for matrices S and T with positive bounded similarity function

Suppose we consider the two binary arrays D and C. h is a function which measures the similarity between the binary description of two entities. For example the Tanimoto coefficient

$$S(i,j)=N(k|d(k,\ell)\wedge d(k,j)=1)/N(k|d(k,i)\vee d(k,j)=1) \qquad (9)$$

Suppose that the coefficients are scaled between 0 and n' and are rounded to the nearest integer. Then $0=<S(i,j),\ T(i,j)=<n'$.

$S(i,j)=0$ indicates complete dissimilarity between objects i and j. $S(i,j)=n'$ indicates that i and j are identical. We now define the matrix A as:-

$$a(\ell,m) = N(w|S(w)=\ell;\ T(w) = m) \tag{10}$$

A is called the incidence matrix for S and T. Denote the first quadrant of (i,j) *wrt* A by $Q'(i,j)$. Let $R(k,\ell)$ be any point in $Q'(i,j)$. Then PR represents a transition of the type (-,+) since

i) $S(p,q)=i$, $T(p,q)=j$ for $a(i,j)$ pairs (p,q)

ii) $S(p',q')=k$, $T(p',q')=\ell$ for $a(k,\ell)$ pairs (p',q')

iii) $i < k$, $j < \ell$ as R is in $Q'(i,j)$

so that $u(p,q:p',q') = d(((s(p,q)-S(p',q'))/T(p,q)-T(p',q'))) = d((i-k)/(j-\ell)) = -1$. Similarly points in $Q''(i,j)$, $Q'''(i,j)$, $Q''''(i,j)$ represent transitions of type (-,-), (+,-) and (+,+), respectively. The ambiguous categories occur when (k,ℓ) lies on the ordinate or abscissa through (i,j). Now consider the point $R(k,\ell)$ again. For each x such that $S(x)=i$ and $T(x)=j$ and for each y such that $S(y)=k$ and $T(y)=\ell$ there is a transition of the type (-,+) between S and T. The contribution to $R(-,+)$ is therefore $a(i,j)a(k,\ell)$ from $R(k,\ell)$. The contribution to $R(-,+)$ from $Q'(i,j)$ is $a(i,j)b(i,j)$ where

$$b(i,j) = \sum_{k=i}^{n'} \sum_{\ell=0}^{j} a(k,\ell) \tag{11}$$

This defines a matrix B such that each element (i,j) is the sum of all the elements (k,ℓ) of A such that $k>i$ and $\ell>j$. B is called the first quadrant accumulant of A and may be derived from A by a simple recurrence relation

$$b(i,j) = a(i,j)+a(i,j-1)+a(i+1,j)-a(i+1,j-1) \quad (12)^*$$

The contribution to $R(-,+)$, however, contains terms for which $i<s$ and $j<t(s)$ and these must be subtracted to obtain expressions for transitions between pairs of objects which have similarity above the respective thresholds. The numbers of transitions in each category are therefore

$$R(-,+) = \sum_{i=s}^{n'} \sum_{j=t(s)}^{n'} (a(i,j)(b(i+1,j-1)-b(i+1,t-1))) \quad (13)$$

$$R(+,-) = \sum_{i=s}^{n'} \sum_{j=t(s)}^{n'} (a(i,j)(b(s,n')-b(i,n')-b(s,j)+b(i,j)))$$

$$R(-,-) = \sum_{i=s}^{n'} \sum_{j=t(s)}^{n'} (a(i,j)(b(i+1,n')-b(i+1,j)))$$

$$R(+,+) = \sum_{i=s}^{n'} \sum_{j=t(s)}^{n'} (a(i,j)(b(s,j-1)-b(s,t-1)-b(i,j-1)+b(i,t-1)))$$

$$R(-,0) = \sum_{i=s}^{n'} \sum_{j=t(s)}^{n'} (a(i,j)(b(i+1,j)-b(i+1,j-1)))$$

$$R(+,0) = \sum_{i=s}^{n'} \sum_{j=t(s)}^{n'} (a(i,j)(b(s,j)-b(i,j)-b(s,j-1)+b(i,j-1)))$$

$$R(0,+) = \sum_{i=s}^{n'} \sum_{j=t(s)}^{n'} (a(i,j)(b(i,j-1)-b(i+1,j-1)-b(i,t-1)+b(i+1,t-1)))$$

$$R(0,-) = \sum_{i=s}^{n'} \sum_{j=t(s)}^{n'} (a(i,j)(b(i,n')-b(i+1,n')-b(i,j)+b(i+1,j)))$$

$$R(0,0) = \sum_{i=s}^{n'} \sum_{j=t(s)}^{n'} (a(i,j)(a(i,j)-1))$$

* The incidence matrix bears a strong similarity to the 'scatter diagram',[4]. However, Kruskal specifically excludes from his diagram the case of repeated occurrence of the same pair (ℓ,m) in the above notation, as he deals with a continuous measure of 'distance' or 'proximity' rather than one which is constrained to take integer values. His subsequent analysis, however, takes care of this point.

The total number of transitions =

$$\sum_{i=s}^{n'} \sum_{j=t(s)}^{n'} (a(i,j)((b(i+1,j)-b(i+1,j-1))-(b(i,j)-b(i,j-1))$$

$$+ a(i,j)+b(s,n')-b(s,t-1)-1) = M(M-1) \text{ where } M = b(s,n')-b(s,t-1)$$

Since the discriminant is symmetrical in $(i,j),(k,1)$ we use the following formulae

$$R(-,s) = \sum_{i=s}^{n'} \sum_{j=t(s)}^{n'} (a(i,j)(b(i+1,j-1)-b(i+1,t-1))) \tag{14}$$

$$R(+,s) = \sum_{i=s}^{n'} \sum_{j=t(s)}^{n'} (a(i,j)(b(i+1,n')-b(i+1,j)))$$

$$R(XE,s) = \sum_{i=s}^{n'} \sum_{j=t(s)}^{n'} (a(i,j)(b(i+1,j)-b(i+1,j-1)))$$

$$R(EX,s) = \sum_{i=s}^{n'} \sum_{j=t(s)}^{n'} (a(i,j)(b(i,n')-b(i+1,n')-b(i,j)+b(i+1,j)))$$

$$R(EE,s) = \sum_{i=s}^{n'} \sum_{j=t(s)}^{n'} (a(i,j)(a(i,j)-1)/2))$$

The total number of transitions is $M(M-1)/2$ in this case. The order of this algorithm is again 2 in n' but n' in this case is independent of the size of the population. For populations smaller than n' this method would be probably less economical than the previous. It may happen, however, that not all of the values of the similarity coefficient between 0 and n' occur. Suppose that only n'' different values occur. The values which occur may be put into correspondence with integers between 0 and n''. The order of the process is then 2 in $n'' < n'$ and must be repeated, unlike the previous one, for each of the n'' thresholds.

The calculation of t is particularly simple for this algorithm. The dependence of t on s is defined in (4). The number of elements in S which have the value s is obtained as the sum of the elements in the s-th column of the incidence matrix.

Similarly the number of elements in T which have the value t is the sum of the elements in the t-th row in the incidence matrix. Thus the number of elements in S which are larger than or equal to s is the sum of the elements in column s onwards. The total number of elements in either S or T is given by the sum of the rows and columns respectively of the incidence matrix. These sums may be obtained from the first quadrant accumulant of the incidence matrix for S and T. The number of elements of S with values larger than or equal to s is $b(s,n')$. The number of elements of T with values larger than or equal to t is $b(1,n')-b(1,t-1)$. The total number of elements in S and T is $b(1,n')$. Therefore, given s, t is chosen to minimise

$$|b(1,n')-b(1,t-1)-b(s,n')| \tag{15}$$

The methods mentioned above were compared by applying them to two matrices with dimension $n = 712$. The matrices were generated from the same binary array consisting of 712 objects described by 200 properties, and were both nine per cent dense. Two similarity coefficients were used. One was the Tanimoto coefficient (see (9)) and the other was the cross correlation coefficient defined by $s(i,j)=(N(k|d(k,i)\wedge d(k,j)=1))/N(k|d(k,i)=1)$ $N(k|d(k,j)=1)$. The purpose of this was to examine the extent to which the similarity between the objects computed on one coefficient agreed with the other. The first method took 3 mins. 50 secs. and the second 14 secs. to calculate all the discrepancies. In the second method this includes the time taken to form the first quadrant accumulant which was 0.63 secs. The discrepancies varied from about 0.96 to 1.00 as the threshold was increased from 20 to 100. The order of the first quadrant accumulant was 60 and the similarity coefficients were scaled between 0 and 100. The number of transitions and the variation of discrepancy with threshold are given in Figs. 4, 5 and 6.

s	size (S)	$t(s)$	size (T)
21	2346	35	2348
22	2291	36	2320
23	2138	40	2157
24	2087	40	2157
25	2086	40	2157
26	1255	50	1284
27	1245	50	1284
28	1200	50	1284
29	1093	50	1284
30	1087	50	1284
31	1051	50	1284
32	1046	50	1284
33	1046	50	1284
34	483	58	425
35	483	58	425
36	477	58	425
37	472	58	425
38	448	58	425
39	444	58	425
40	373	63	387
41	373	63	387
42	372	63	387
43	353	64	351
44	353	64	351
45	345	66	346
46	342	66	346
47	341	66	346
48	341	66	346
49	341	66	346
50	341	66	346
51	113	72	112
52	113	72	112
53	113	72	112
54	113	72	112
55	112	72	112
56	109	73	109
57	109	73	109
58	106	75	107
59	106	75	107
60	106	75	107
61	99	78	99
62	97	81	96
63	97	81	96
64	97	81	96
65	96	81	96
66	96	81	96
67	68	86	68
69	68	86	68
70	68	86	68
72	68	86	68
73	68	86	68
75	68	86	68
77	64	89	64
78	64	89	64
80	64	89	64
81	60	100	60
86	60	100	60
89	60	100	60
100	60	100	60

Fig.4 Comparable thresholds applied to the two matrices

THRESHOLD	*-R*	*XE*	*+R*	*EX*	*EE*
20	101331	106551	2005694	259732	294403
21	101275	106427	1987106	259720	294387
22	93277	105676	1869850	258670	293952
23	71509	87390	1507015	232977	290710
24	56241	83654	1468171	232709	290600
25	56241	83654	1468171	232709	290600
26	20695	9148	375076	84164	103742
27	20695	9148	375076	84164	103742
28	19903	9142	373696	84164	103741
29	10327	8296	357928	84164	103465
30	9953	8905	357259	84164	103465
31	5899	6269	346150	84065	103428
32	5532	6249	345491	84065	103428
33	5532	6249	345491	84065	103428
34	908	514	59711	5448	21749
35	908	514	59711	5448	21749
36	906	513	59290	5448	21749
37	906	513	59290	5448	21749
38	786	369	54610	5448	21683
39	744	366	53428	5446	21682
40	32	278	33735	5259	21054
41	32	278	33735	5259	21054
42	30	276	33387	5259	21054
43	22	276	32007	5251	21052
44	22	276	32007	5251	21052
45	14	220	30011	5251	21045
46	14	220	30011	5251	21045
47	0	210	29694	5251	21045
48	0	210	29694	5251	21045
49	0	210	29694	5251	21045
50	0	210	29694	5251	21045
51	0	7	4021	12	406
52	0	7	4021	12	406
53	0	7	4021	12	406
54	0	7	4021	12	406
55	0	7	4021	12	406
56	0	7	3694	12	403
57	0	7	3694	12	403
58	0	2	3381	10	402
59	0	2	3381	10	402
60	0	2	3381	10	402
61	0	2	2688	0	391
62	0	0	2400	0	390
63	0	0	2400	0	390
64	0	0	2400	0	390
65	0	0	2400	0	390
66	0	0	2400	0	390
67	0	0	496	0	12
69	0	0	496	0	12
70	0	0	496	0	12
72	0	0	496	0	12
73	0	0	496	0	12
75	0	0	496	0	12
77	0	0	240	0	6
78	0	0	240	0	6
80	0	0	240	0	6

Fig. 5 Number of transitions in each of the categories

ASSIGNMENT OF AMBIGUOUS CATEGORIES

	1 TO REVERSALS			2 TO NONREVERSALS		
s	$R(-,s)$	$R(+,s)$	$g(-,s)$	$R(-,s)$	$R(+,s)$	$g(+,s)$
20	762017	2005694	72.47	101331	2666380	96.34
21	761809	1987106	72.29	101275	2647640	96.32
22	751575	1869850	71.33	93277	2528148	96.44
23	682586	1507015	68.83	71509	2118092	96.73
24	663204	1468171	68.88	56241	2075134	97.36
25	663204	1468171	68.88	56241	2075134	97.36
26	217749	375076	63.27	20695	572130	96.51
27	217749	375076	63.27	20695	572130	96.51
28	216950	373696	63.27	19903	570743	96.63
29	206882	357928	63.37	10327	554483	98.17
30	206487	357259	63.37	9953	553793	98.23
31	199661	346150	63.42	5899	539912	98.92
32	199274	345491	63.42	5532	539233	98.98
33	199274	345491	63.42	5532	539233	98.98
34	28619	59711	67.60	908	87422	98.97
35	28619	59711	67.60	908	87422	98.97
36	28616	59290	67.45	906	87000	98.97
37	28616	59290	67.45	906	87000	98.97
38	28286	54610	65.88	786	82110	99.05
39	28238	53428	65.42	744	80922	99.09
40	26623	33735	55.89	32	60326	99.95
41	26623	33735	55.89	32	60326	99.95
42	26619	33387	55.64	30	59976	99.95
43	26601	32007	54.61	22	58586	99.96
44	26601	32007	54.61	22	58586	99.96
45	26530	30011	53.08	14	56527	99.98
46	26530	30011	53.08	14	56527	99.98
47	26506	29694	52.84	0	56200	100.00
48	26506	29694	52.84	0	56200	100.00
49	26506	29694	52.84	0	56200	100.00
50	26506	29694	52.84	0	56200	100.00
51	425	4021	90.44	0	4446	100.00
52	425	4021	90.44	0	4446	100.00
53	425	4021	90.44	0	4446	100.00
54	425	4021	90.44	0	4446	100.00
55	425	4021	90.44	0	4446	100.00
56	422	3694	89.75	0	4116	100.00
57	422	3694	89.75	0	4116	100.00
58	414	3381	89.09	0	3795	100.00
59	414	3381	89.09	0	3795	100.00
60	414	3381	89.09	0	3795	100.00
61	393	2688	87.24	0	3081	100.00
62	390	2400	86.02	0	2790	100.00
63	390	2400	86.02	0	2790	100.00
64	390	2400	86.02	0	2790	100.00
65	390	2400	86.02	0	2790	100.00
66	390	2400	86.02	0	2790	100.00
67	12	496	97.64	0	508	100.00
69	12	496	97.64	0	508	100.00
70	12	496	97.64	0	508	100.00
72	12	496	97.64	0	508	100.00
73	12	496	97.64	0	508	100.00
75	12	496	97.64	0	508	100.00
77	6	240	97.56	0	246	100.00
78	6	240	97.56	0	246	100.00
80	6	240	97.56	0	246	100.00

Fig. 6 Variation of discrepancy with threshold

5. General remarks about the dependence of the discrepancy g on the threshold

In the preceding paragraphs it was suggested that the discrepancy may be used to examine the relationship between two symmetrical matrices of the same order. It is now necessary to clarify this remark by establishing what are the overall characteristics of $g(s)$ which enable the relationship to be described.

At present we are interested in matrices generated from binary arrays by applying a similarity function. The similarity function induces a partial ordering on pairs of points (i,j). The discrepancy measures the extent to which the portions of the partial orderings generated in this way agree with each other above a particular level.

Let us suppose that the two matrices are generated from the same binary object-property array, by applying two different similarity functions. Since the functions are intended to measure the resemblance between pairs of points, the partial orderings are expected to agree with each other to a large extent. In general it is to be expected that pairs of objects with high similarity with respect to maximum similarity in the respective matrices will be ordered identically. In the lower ranges of similarity, the different treatment of objects by the two functions is expected to become apparent, with the result that the partial orderings may differ considerably. Thus the discrepancy is expected to be near its maximum value for high thresholds and to decrease as the threshold is reduced. The discrepancy is therefore expected to be monotonically increasing with respect to the threshold. At low thresholds the discrepancy should still be far from its minimum value, indicating that even at low values the similarity do not disagree completely. The observed behaviour of the discrepancy as tabulated in Fig. 6 may be seen to agree with the expected behaviour.

The above example was constructed to confirm that the expected behaviour of the discrepancy did indeed occur in practice.

Suppose now that we take two binary arrays, one being a classification of the other, and that we proceed to evaluate the discrepancies of the matrices derived from these arrays by applying the Tanimoto similarity function to each. The same general behaviour as in the previous case is observed. The discrepancy is a monotonic increasing function of the threshold. The minimum value of the discrepancy, however, lies below the corresponding value for the previous case. Thus, among the lower similarities, the classification is less well ordered with respect to the data. Some intuitive justification for this is provided by remarking that classification is an irreversible process in that the original information about the individual objects of the population is not completely recoverable from the classification. The discrepancy gives an indication of the extent to which the ordering of the objects among themselves is recoverable in spite of the fact that some of the information about the objects has been lost. The value of the minimum discrepancy gives the maximum departure from complete recoverability of the ordering. The largest non-zero value of the threshold is nearly the same as in the previous case. This is the largest similarity for which there is any connection between at least two pairs of objects in the two arrays.

A last example is provided by the comparison of the original data array, C, with a randomly generated array. Briefly, the random array was generated to provide a random description of the population of objects using the same set of attributes. A random number of randomly selected attributes was assigned to each object. The maximum number of attributes assigned to any object was arranged to be the same as for the data array. The dependence of the discrepancy on the threshold is no longer monotonic and the values of $\min(g(s))$ and $\max(s)$ for each of the three cases mentioned. Fig. 8 gives the discrepancy curve for the last two examples. Fig. 7 provides sufficient evidence for a distinction to be made between each of the cases described. The times taken to evaluate the discrepancy for all thresholds are included for interest.

ARRAY	SIMILARITY	min($g(s)$)	max(s)	MONOTONICITY	TIME(SECS)
data	Cross Corr	0.96	80	increasing	14
clsf	Tanimoto	0.75	71	increasing	14
rdnm	Tanimoto	0.60	25	non-monotone	6

Fig. 7 Comparison of the characteristics of the discrepancy for the data array (using Tanimoto similarity) with respect to each of the indicated arrays

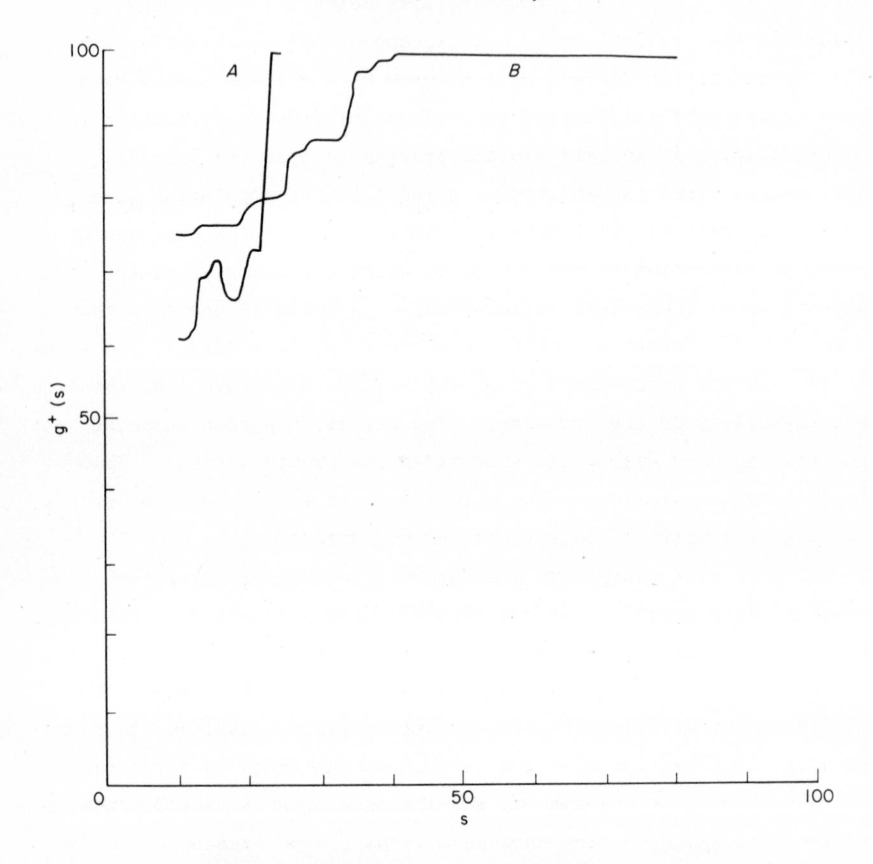

Fig. 8. Graph of discrepancy against threshold for:-

A. Random data array
B. A classification array

The partial orderings are induced by applying the Tanimoto similarity function and are compared with the partial ordering induced by the same function on the data array.

Acknowledgements

My thanks are due to Professor M.V. Wilkes and the Staff of the Cambridge University Mathematical Laboratory on whose computer this work was carried out. This work was supported in part by the U.K. Office for Scientific and Technical Information.

References

1. Sparck Jones, K., and Jackson, D.M. The Use of the Theory of Clumps for Information Retrieval. *Cambridge Language Research Unit Report 200*, mimeo, June, 1967.
2. Ball, G.H. Data Analysis in the Social Sciences: What about the details. *Proceedings of the Fall Joint Computer Conference, 1965.*
3. Kendall, M.G. The Advanced Theory of Statistics, Vol. 1, C. Griffin and Co. Ltd., London, 1945.
4. Kruskal, J.B. Multidimensional scaling by optimising goodness of fit to a non-metric hypothesis. *Psychometrika, 29*, 1964.

Discussion

Q.(Cole) Can you tell us something of how you propose to use these ideas in your projected algorithm?

A. My main purpose in formulating this method was to enable me to compare classifications. I then had the idea that one could determine what type of curve one would like to get. I have the feeling that it will lead towards multidimensional scaling. Basically, my criteria are ones of producing a more economical description which would correspond to the one of finding the minimum dimension. The effect that one wants the functions $g(+,s)$ and $g(-,s)$ to be monotonic, breaks down to the monotonicity of the similarity measure against "proximity".

Q. (Parker-Rhodes) Is there a simple way in which one can calculate $R(+,s)$ for one particular value of s and then use this in order to calculate it for a slightly different value of s without going through the whole process again?

A. As far as I can see one would have to have very detailed information about the distribution. It is particular information of this kind that I have used in the algorithm. I suspect, therefore, that if one doesn't want to calculate an estimate then one is forced to go through these processes one by one.

Q. (Parker-Rhodes) As a user taxonomist I can appreciate that there is a need for a method of comparing classifications.

A. Yes, I have formed the same opinion.

Q. (Parker-Rhodes) In principle it would seem that one can use this kind of algorithm as a means of assessing the excellence of a classification in relation to the original data. If you get a nice straight line at the end you've won and if it's a "wiggly" one then that's very bad. On the face of it, this seems too good to be true. Can you comment on this?

A. This is a difficult question to answer. Certainly if the line is straight then one can make use of the conclusions I specified and I work on the hypothesis that it is those assumptions - namely of being able to make inferences about the population of objects from a knowledge of the classifications - that one wants to make. One is at liberty to say that this kind of thing doesn't matter in which case I suggest that you don't use this algorithm: but if you _do_ use it then you are forced by the analysis to say that the straight line is more or less what one wants. I can also explain the monotonic relationship between the two but that would take some time to go in to. There are obvious intuitive reasons for seeing why this should be the case. It also enables one to distinguish between an actual classification - that is something one can put through a sort of classification engine - and random data. If one could not distinguish these it would be a measure of one's failure.

ON TWO CRITERIA OF CLASSIFICATION

I. C. Lerman

Centre de Calcul, Maison des Sciences de l'Homme

I. Introduction

Let E be an N-set of objects $x,y,z,\ldots$; let F denote the set of unordered object pairs: $F = \{(x,y), x\in E, y\in E, x\neq y\}$. The basic data is a pre-order on F. $p = (x,y)$ precedes $q = (z,t)$ if x and y are less similar than z and t. Let ω denote the graph of pre-order relation, ω is a subset of the product $F \times F$.

A partition of E determines a pre-order on F with two classes R and S; (x,y) belongs to R if and only if x and y are in the same class of the partition, S is the complement of R in F; that is to say; (z,t) belongs to S if and only if z and t belong to distinct classes, $F=R+S$. Any pair of S precedes any pair of R.

One of the essential purposes of taxonomy is to find one partition of E which "best approximates" the similarities of objects. The degree of the similarity between two objects is expected to be high if these belong to the same class and low if not. More precisely, at the beginning we define a numerical function which measures the accordance between a classification (*e.g.* partition) and the graph ω; this function is called <u>criterion</u>. In fact the criterion will be a mapping of Π into N, where Π denotes the set of all partitions of E. Then the problem is to find the partitions which maximise the value of the criterion.

1) J.P. Benzecri proposed as a criterion the cardinal of the intersection of ω and $S \times R$, in $F \times F$.

$$P\in\Pi \xrightarrow{a_1} a_1\ (P) = |\omega\ S \times R|$$

2) W.F. de la Vega proposed as a criterion the cardinal of the complement of the symmetric difference between ω and $S \times R$, in $F \times F$:

$$P \in \Pi \xrightarrow{a_2} a_2\,(P) = |F \times F - \omega \Delta\, S \times R|$$

Considering the relations:

$$|\omega \cup S \times R| = |\omega \Delta S \times R| + |\omega \cap S \times R| = |\omega| + |S \times R| - |\omega \cap S \times R|$$

We obtain: $a_2\,(P) = |F|^2 - |\omega| + 2(\omega \cap S \times R| - \frac{1}{2}|S| \times |R|)$

Then the optimizing problem consists in finding classifications which maximise $|\omega \cap S \times R|$ or $|\omega \cap S \times R| - \frac{1}{2}|S| \times |R|$ accordingly to the first criterion or to the second one.

Our purpose is to compare these two criteria.

II. The theorem of two criteria comparisons

II.A. Preliminaries: On some enumerative problems connected with the partitions of E.

Definitions and Generalities: We call the decreasing sequence of cardinals of the different classes the partition type.

So, $(n_1, n_2, n_3, ---, n_i, ---)$; $n_1 \geq n_2 \geq --- n_i \geq ---$ is the type of one partition of E for which the cardinals of the classes are respectively in the decreasing order $n_1, n_2, ---, n_i$ ---.

$N = |E| = \sum_{i=1}^{\infty} n_i$. The partition of E has exactly k classes if the last subscript such as $n_i \neq 0$, is k. A subjective mapping f of E into the collection of numbers $(1, 2, ---, k)$ determines a partition of E with k classes to each of which a number is assigned. f will be called "a partition with labelled classes" or more briefly a labelled partition. Let $\mathfrak{F}$ denote the set of the partitions with labelled classes the type of which $(n_1, n_2, ---, n_k)$ is given; if f belongs to $\mathfrak{F}$, $f^{-1}(i) = n_i$ where $f^{-1}(i)$ denotes the class of number i, that can be also noted E_i.

On the other hand, let Π be the set of the "partitions with unlabelled classes" the type of which $(n_1, n_2, ---, n_k)$, is given. The cardinal of $\mathfrak{F}$ is given by the following formula

$$|\mathfrak{F}| = \frac{N!}{n_1!\, n_2! \cdots n_k!}$$

Generally, the integers $n_1, n_2, \cdots, n_k$ are not mutually distinct. If a partition ω of Π has k_1 classes with the same cardinal $\nu_1, \ldots, k_2$ classes with the same cardinal $\nu_2, \cdots$, and k_r classes with the same cardinal ν_r ($k_1 + k_2 + \cdots + k_r = k$ and $k_1\nu_1 + k_2\nu_2 + \cdots + k_r\nu_r = N$) we have:

$$|\Pi| = \frac{N!}{k_1!\, k_2! \cdots k_n!\, n_1!\, n_2! \cdots n_k!}$$

For one element P of Π the cardinals of R and S are given by the following formulae

$$|R| = \sum_{i=1}^{k} n_i\,(n_i - 1)/2 \text{ and } |S| = \sum_{i<j} n_i n_j \;;$$

on the other hand $|R| + |S| = |F| = N(N-1)/2$. Then $|R|$ determines $|S|$.

2 - Lemmas

2.1 The proportion in $\mathfrak{F}$ of elements for which a given pair (x,y) is contained in the same class is $\sum_{i=1}^{k} n_i\,(n_i-1)/N(N-1)$.

Our purpose is to determine the cardinal of the following subset of $\mathfrak{F}$ $\{ f \varepsilon \mathfrak{F} \mid f(x) = f(y) \}$; which may be expressed as a sum of subsets:

$$\{ f \varepsilon \mathfrak{F} \mid f(x) = f(y)\} = \sum_{i=1}^{k} \{ f \varepsilon \mathfrak{F} \mid f(x) = f(y) = i \}$$

Hence $$|\{ f \varepsilon \mathfrak{F} \mid f(x) = f(y)\}| = \sum_{i=1}^{k} |\{ f \varepsilon \mathfrak{F} \mid f(x) = f(y) = i \}|$$

The subset $\{f(x)=f(y)=i\}$ can be mapped 1 - 1 onto the set of labelled partitions of $E - \{x,y\}$; such that the cardinals of different classes of one partition are respectively

$$n_1, n_2, ---, n_{(i-1)}, n_i - 2, n_{(i+1)}, ---, n_k$$

subject to $n_i \geq 2$

Therefore $|\{f(x) = f(y) = i\}| = (N-2)!/n_1!n_2! --- (n_i-2)!--n_k!$ if $n_i \geq 2$ and 0 if not

Hence

$$|\{ f(x) = f(y)\}| = \sum_{\{i,n_i \geq 2\}} (N-2)!/n_1!n_2! --- (n_i-2)!---n_k!$$

The consideration of the ratio of this cardinal on $|\mathfrak{F}|$ leads to the result above.

We abbreviate the following demonstrations which are analogous to this one.

2.2 The proportion in $\mathfrak{F}$ of elements, for which two objects of a given pair (x,y), belong, respectively, to distinct classes, is,

$$2 \sum_{i<j} n_i n_j / N(N-1)$$

$$|\{f \varepsilon \mathfrak{F} \; |f(x) \neq f(y)\}| = \sum_{i \neq j} |\{ f \varepsilon \mathfrak{F} \; |f(x)=i, \; f(y) = j \}|$$

The sum includes $k(k-1)$ terms.

The set $\{f(x) = i, f(y) = j, i \neq j\}$; can be mapped 1 - 1 onto the set of "labelled partitions classes" of $E - (x,y)$; such that the cardinals of classes of one partition are respectively:

$$n_1, n_2, ---, n_{(i-1)}, n_i - 1, n_{i+1}, ---, n_{(j-1)}, n_j - 1, n_{j+1}, ---, n_k$$

This remark is sufficient to find the result.

2.3 Let x, y and z be three given objects of E, the proportion in $\mathfrak{F}$ of elements for which x and y are in the same class, x and z in distinct classes is $\sum_{i \neq j} n_i(n_i-1)n_j/N(N-1)(N-2)$

$$|\{f \in \mathfrak{F} \mid f(x)=f(y) \neq f(z)\}| = \sum_{H} |\{f \in \mathfrak{F} \mid f(x)=f(y)=i,\ f(z)=j\}|$$

where H is $\{(i,j) \mid i \neq j,\ n_i \geq 2\}$.

The set $\{f \in \mathfrak{F} \mid f(x) = f(y) = i,\ f(z) = j,\ i \neq j$ and $n_i \geq 2\}$ can be mapped 1 - 1 onto the set of "labelled partitions of $E - (x,y,z)$ so that the cardinals of different classes subject to $n_i \geq 2$, are respectively

$$n_1, n_2, ---, n_{(i-1)}, n_i-2, n_{(i+1)}, ---, n_{(j-1)} n_j-1, n_{j+1}, ---, n_k.$$

It is now easy to express the current term of the upper sum and to establish the result.

2.4 Let x, y, z and t, be four objects of E, the proportion in $\mathfrak{F}$ of elements for which x and y are in a same class, z and t in distinct classes is

$$\left[\sum_{J} n_i(n_i-1)n_j n_\ell + 2 \sum_{k} n_i(n_i-1)(n_i-2)n_j\right]/N(N-1)(N-2)(N-3)$$

where $J = \{(i,j,\ell) \mid i \neq j,\ j \neq \ell,\ \ell \neq i,\ n_i \geq 2\}$, $K = \{(i,j) \mid i \neq j,\ n_i \geq 3\}$

The first sum involves $k(k-1)(k-2)$ terms and the second one $k(k-1)$ terms.

An element f of $\mathfrak{F}$ satisfies $f(x) = f(y)$ and $f(z) \neq f(t)$ if and only if f answers to one of the following alternatives.

1. $f(x) = f(y) = i$, $f(z) = j$ and $f(t) = \ell$ for an ordered triplet (i,j,t) such as $i \neq j$, $j \neq \ell$, $\ell \neq i$ and $n_i \geq 2$.

or

2.a $f(x) = f(y) = f(z) = i$ and $f(t) = j$

or

b $f(x) = f(y) = f(t) = i$ and $f(z) = j$

for a couple (i,j) such as $i \neq j$ and $n_i \geq 3$.
The cardinal of the subset of $\mathcal{F}$ whose elements satisfy 1 is the one of the set of labelled partitions of $E - (x,y,z,t)$, such as the cardinals of classes are respectively:

$$n_1, n_2, ---, n_{(i-1)}, n_i-2, n_{(i+1)}, ---, n_{(j-1)}, n_j-1, n_{j+1}, ---,$$
$$n_{(\ell-1)}, n_\ell-1, n_{\ell+1}, ---, n_k.$$

The cardinal of the subset of $\mathcal{F}$ whose elements satisfy 2.a (resp. 2.b) is the same as the set of labelled partitions of $E - (x,y,z,t)$ such that the class cardinals are respectively

$$n_1, n_2, ---, n_{(i-1)}, n_i-3, n_{i+1}, ---, n_{(j-1)}, n_j-1, n_{j+1}, ---, n_k.$$

With these remarks we shall complete the calculation.

2.5 The different ratios, that we have just derived still hold if we consider them in Π instead of in $\mathcal{F}$.
To one element P of Π corresponds, exactly, $k_1!\, k_2!\, ---\, k_r!$ elements of $\mathcal{F}$ by labelling in $k_i!$ ways the k_i classes with the same cardinal ν_i, $i = 1, 2, ---, r$ [cf. IIA.1]. Let us suppose that we have calculated one of the previous proportions (1,2,3 or 4) in Π; we obtain the same proportion in $\mathcal{F}$ by multiplying the two terms of the ratio into $k_2! \times k_2! \times --- \times k_r!$. Because one element P of Π intervenes in the enumeration of a given term of the ratio if and only if the corresponding $k_1!\, k_2!\, ---\, k_r!$ elements in $\mathcal{F}$ intervene in the enumeration of this term of the ratio defined in $\mathcal{F}$. Henceforth we shall work in Π.

2.6 Important Remarks

1. For reasons of symmetry, the different proportions which had been defined, hold whatever the objects that form the pair or the couple of pairs.

2.a. Let G denote the set of couples of pairs such that the two pairs have one common component:

$G = \{((x,y), (x,z))$ where x,y and z are mutually distinct$\}$.

The cardinal of G is $N(N-1)\ (N-2)$; $N = |E|$

b. Let H denote the set of couples of pairs such that the two pairs have no common component.

$H - \{((x,y), (z,t))$ where x,y,z and t are mutually distinct$\}$.

The cardinal of H is $N(N-1)\ (N-2)\ (N-3)/4$.

II. B. Theorem. If ω is a total order on F, the mean in Π of $|\omega \cap S \times R| - \frac{1}{2}|S| \times |R|$ is null.

If P is an element of Π, there corresponds to it R and S for which we have:

$$\frac{1}{2}|S| \times |R| = \frac{1}{4}\left[\sum_{i=1}^{k} n_i(n_i-1)\right] \times \left[\sum_{i<j} n_i n_j\right] \qquad (1)\,[\text{cf.}\ \S\text{IIA.1}]$$

We denote:

$\Phi[(p,q), S \times R]$ the indicator function of the set $S \times R$:

$\Phi[(p,q), S \times R] = 1$ if $(p,q) \in S \times R$ and 0 if not

$\Phi[(p,q), \omega]$ the indicator function of ω.

The mean in Π of $|\omega \cap S \times R|$ may be written as follows:

$$\frac{1}{|\Pi|} \sum_{P \in \Pi} \sum_{(p,q) \in F \times F} \Phi\ [(p,q), S \times R]\Phi[(p,q), \omega] \quad (2)$$

Let us decompose the sum over $F \times F$ into two sums: The former will be over G and the latter of H. [cf. the above remarks], namely:

$$\frac{1}{|\Pi|} \sum_{P \in \Pi} \sum_{(p,q) \in G} \Phi[(p,q), S \times R]\Phi[(p,q),\omega] \quad +$$

$$\frac{1}{|\Pi|} \sum_{P \in \Pi} \sum_{(p,q) \in H} \Phi[(p,q), S \times R]\Phi[(p,q),\omega] \qquad (3)$$

If we reverse the sum signs, the expression (3) may be written as

$$\sum_G \Phi[(p,q),\ \omega] \times \frac{1}{|\Pi|} \sum_\Pi \Phi[(p,q),\ S \times R] + \sum_H \Phi[(p,q),\ \omega] \times$$

$$\frac{1}{|\Pi|} \sum_\Pi \Phi[(p,q),\ S \times R] \tag{4}$$

ω being a total order: $(p,q)\varepsilon\omega \leftrightarrow (q,p)\varepsilon\omega$.

On the other hand: $(p,q)\varepsilon G \rightleftarrows (q,p)\varepsilon G$ and $(p,q)\varepsilon H \leftrightarrow (q,p)\varepsilon H$.

By the lemmas (3) and (4) and according to the above remarks, the expression (4) becomes:

$$\frac{1}{2} N(N-1)(N-2) \Sigma\ n_i(n_i-1)n_j/N(N-1)(N-2)$$

$$+ \frac{1}{8} N(N-1)(N-2)(N-3) [\sum_J n_i(n_i-1)n_j n_\ell + 2 \sum_{i \neq j} n_i(n_i-1)(n_i-2)n_j] /$$

$$N(N-1)(N-2)(N-3) \tag{5}$$

where $J = \{(i,j,\ell), i \neq j,\ j \neq \ell,\ \ell \neq i$ and $n_i \geq 2\}$.

The expression (5) becomes

$$\frac{1}{2} [\sum_{i \neq j} n_i(n_i-1)n_j + \frac{1}{4} (\sum_J n_i(n_i-1)n_j n_\ell + 2 \sum_{i \neq j} n_i(n_i-1)(n_i-2)n_j)]$$

$$\tag{6}$$

The last sum, from left to right, may be expressed as follows

$$\sum_{i \neq j} n_i^2(n_i-1)n_j - 2 \sum_{i \neq j} n_i(n_i-1)n_j\ .$$

Simplifying, the expression (6) we obtain:

$$[\frac{1}{2}[\frac{1}{4}\sum_{i=1}^{k} n_i(n_i-1)(\sum_I n_j n_\ell + 2n_i \sum_{j\neq i} n_j]$$ where $I = \{(j,\ell),\ j\neq i \text{ and } \ell\neq i\}$.

We have: $\sum_I n_j n_\ell + 2n_i \sum_{j\neq i} n_j = \sum_{r\neq s} n_r n_s$

So, the assertion is shown.

III. Type of partition and cardinal of the corresponding set R.

1) Introduction

Remembering that if $(n_1, n_2, ---, n_i, ---)$, where $n_1 \geq n_2 \geq --- \geq n_i \geq ---$, is the type of partition of E, for which there corresponds R and S; we have $|R| = \sum_i n_i(n_i-1)/2 = \frac{1}{2}(\sum_i n_i^2 - N)$ where $N = |E|$.

$|S| = \sum_{i<j} n_i n_j$ is the complement to $|F|$ of $|R|$; $|F| = N(N-1)/2$.
It can be easily established that in the set of all partitions of E into two classes, $|R|$, or equivalently Σn_i^2, determines univocally the type of the partition. This elementary result leads us to deal with the following problem: If Σn_i^2 is given as equal to a positive integer M, then the set of partition-types such as $\Sigma n_i^2 = M$ is to be determined. We resolved this problem by a recursive technique. We shall not speak of this problem; we shall denote by $\Psi(N,M)$ the set of partition-types for which: $\Sigma n_i = N$ and $\Sigma n_i^2 = M$, N and M being given.

2) Proposition

The partitions which maximize $|R \times S|$ are those for which $|R|$ is the closest of $N(N-1)/4$. This result is evident. $|R \times S| = |R| \times |S|$ is the product of two cardinals whose sum is fixed; $|R| + |S| = N(N-1)/2$. If $\Psi(N,N(N+1)/2)$ is different from the empty set, there exists, at least one type of partition for which

$|R|$ is equal to $N(N-1)/4$. For example, if N is the square of an integer $(N + \sqrt{N})/2$, $(N - \sqrt{N})/2)$ is the type of a partition which realises $|R| = N(N-1)/4$; beforehand it should be noted that $(N + \sqrt{N})/2$ and $(N - \sqrt{N})/2$ are integers since N and $\sqrt{N}$ have the same parity.

IV On some aspects of comparison of the two criteria

The following propositions are corollaries of the main theorem of paragraph II. Let $\mathcal{R}$ denote the set of partitions for which $|R| = r$, where r is a given integer; this set is not empty if and only if $\Psi(N, 2r + N)$ is not empty.

1. Proposition 1 If ω is a total order on F, the mean in $\mathcal{R}$ of $|\omega \cap S \times R| - \frac{1}{2}|S| \times |R|$ is null.

The sets Π_t, where Π_t is the set of partitions of type t belonging to $\Psi(N, 2r + N)$, form a partition of $\mathcal{R}$:

$$\mathcal{R} = \sum_{t \in T} \Pi_t \text{ (sum of sets), where } T = \Psi(N, 2r + N)$$

The mean of $|\omega \cap S \times R| - \frac{1}{2}|S| \times |R|$ in each set Π_t is null [cf. §II.B]; then the mean of the same quantity in $\mathcal{R}$ is also null.

2. Proposition 2 The two criteria are equivalent in $\mathcal{R}$. This follows from the fact that the difference between the two criteria depends uniquely on N and r.

3. Proposition 3 If ω is a total order on F and the cardinal N of E such as $\Psi(N, N(N+1)/2)$ different from the empty set, then a partition for which

$$|R| < \frac{N(N-1)}{4}\left(1 - \frac{\sqrt{2}}{2}\right) \text{ or } |R| > \frac{N(N-1)}{4}\left(1 + \frac{\sqrt{2}}{2}\right)$$

can not be optimal according to the first criterion.

Since $\Psi(N,N(N+1)/2)$ is not empty, there is at least one partition ω for which $|S| \times |R|$ equals $N^2(N-1)^2/2^4$. [cf. III.2]. If t is an arbitrary element of $\Psi(N,N(N+1)/2)$, by applying the theorem II.B. there exists in Π_t at least one partition P_t for which:

$$|\omega \cap S_t \times R_t| - \frac{1}{2} |S_t| \times |R_t| \geq 0$$

that is to say

$$|\omega \cap S_t \times R_t| \geq \frac{1}{2^5} N^2(N-1)^2 \tag{1}$$

where S_t and R_t are referred to P_t.

P_m denoting an optimal partition for the first criterion (maximum of $|\omega \cap S \times R|$), S_m and R_m being related to P_m, we have:

$$|\omega \cap S_m \times R_m| > |\omega \cap S_t \times R_t| > \frac{1}{2^5} N^2(N-1)^2 \tag{2}$$

But $|S_m \times R_m| \geq |\omega \cap S_m \times R_m|$, then necessarily

$$|S_m \times R_m| > \frac{1}{2^5} N^2(N-1)^2 \tag{3}$$

In other words a partition for which

$$|S \times R| < \frac{1}{2^5} N^2(N-1)^2 \tag{4}$$

can not be optimal for the first criterion.

We have: $|S \times R| = |S| \times |R| = [N(N-1)/2 - |R|] \times |R|$.

Putting $|R| = r$ and $N(N-1)/2 = s$, the inequality (4) becomes

$$r(s-r) < \frac{s^2}{2^3}$$

which holds if and only if

$$r < \frac{s}{2}\left(1 - \frac{\sqrt{2}}{2}\right) \quad \text{or} \quad r > \frac{s}{2}\left(1 + \frac{\sqrt{2}}{2}\right) \tag{5}$$

The condition $\Psi(N,N(N+1)/2) \neq \Phi$ is not a restrictive one, if N is an arbitrary integer, a similar result can be obtained by replacing the right member of (1) by the maximum value of $\frac{1}{2}$ $|S| \times |R|$, in the set of all partitions of E.

4. Proposition 4 The integer N is assumed to satisfy the two following conditions: 1) $\Psi(N,N(N+1)/2)$
2) N is divisible by k: $N = n.k$.

If $k > 6$ or $k = 1$, a partition of E, into k classes with the same cardinal n, cannot be optimal for the first criterion.

If P is a partition of E into k classes with the same cardinal n, we have

$$|R| = kn(n-1)/2 \tag{1}$$

where R is associated to ω.

According to the above proposition, a partition for which $|R| < \frac{N(N-1)}{4}(1 - \frac{\sqrt{2}}{2})$, cannot be optimal for the first criterion. Therefore if

$$kn\,(n-1)/2 < \frac{kn(kn-1)}{4}\left(1 - \frac{\sqrt{2}}{2}\right) \tag{2}$$

the partition P cannot be optimal for the first criterion.
The inequality (2) can be written

$$k > 4 + 2\sqrt{2} - \frac{5 + 2\sqrt{2}}{n} \tag{3}$$

(3) is necessarily satisfied for $k > 6$.

On the other hand, according to the above proposition a partition for which $|R| > \frac{N(N-1)}{4}(1 + \frac{\sqrt{2}}{2})$ cannot be optimal for the first criterion. Therefore if

$$\frac{kn(n-1)}{2} > \frac{kn(kn-1)}{4}\left(1 + \frac{\sqrt{2}}{2}\right) \tag{4}$$

the partition P cannot be optimal.

Simplifying the inequality (4) we obtain

$$k < 2(2 - \sqrt{2}) - \frac{3 - 2\sqrt{2}}{n}$$

This inequality is necessarily satisfied for $k = 1$ and $N > 1$.

We illustrate the last proposition with an elementary example: For $N = 49$, the set $\Psi(N, N(N+1)/2) = \Psi(49, 1225)$ is not empty; effectively N is a square of an integer, the type $(N + \sqrt{N})/2$, $(N - \sqrt{N})/2) = (28, 21)$ belongs to $\Psi(49, 1225)$. By the preceding proposition a classification of a set E, whose cardinal is 49, into 7 classes with the same cardinal 7 cannot be optimal for the first criterion.

5) Conclusion We have shown that the two criteria are equivalent in a set of partitions for which $|R|$ is constant. In a set of partitions of given type, the mean of the second criterion is independent of this type. This proposition shows that the second criterion has an intrinsic character. On the other hand, we have shown (proposition 3) and illustrated (proposition 4) an impossibility for some partitions to be optimal for the first criterion. In other words the first criterion is biased. For this reason the second criterion is preferable to the first one.

References

J.P. Benzecri 1) Problemes et methodes de la taxonomie, Rennes (1965b) et cours I.S.U.P., Paris 1965-1966. 2) Ordre lateral entre lois de probabilite sur un ensemble ordonne; applications a l'Analyse Factorielle et aux criteres de classification (I.S.U.P. Paris 1968).

P. Dagnelie, A propos des differentes methodes de classification numerique - *Revue de Statistique Appliquee* - (I.S.U.P.) 1966 - Vol. XIV no.3.

I.C. Lerman 1) Indice de Similarite et Preordonnance associee. Communication a Seminaix, Centre de Calcul - Maison des Sciences de l'Homme - Paris 1967. 2) Sur deux criteres de la classification. Centre de Calcul - M.S.H. Paris 1967. 3) Analyse de probleme de la recherche d'une hierarchie de classifications. Centre de Calcul - M.S.H. Paris 1968.

S. Regnier, Sur quelques aspects mathematiques des problemes de classification automatique. Centre de Calcul - M.S.H. Paris 1965 et revue I.C.C. 1965. Vol. 4 pp. 175-191.

R.N. Shepard, The analysis of proximities: scaling with an unknown distance function, I & II, *Psychometrika* 1962.

R.R. Sokal et P.M.A. Sneath, Principles of numerical taxonomy, W.H. Freeman and Cie, San Francisco and London; 1963.

W.F. de la Vega, Techniques de classification automatique utilisant un indice de ressemblance. Centre de Calcul- M.S.H. Paris 1965. p.48-54.

Discussion

Q. (Jackson) In minimising your criterion function do you examine each of the very many possible partitions or have you certain heuristic methods which you haven't mentioned?

A. I said, we obtain only local optima and not a general optimum. We can't examine all possibilities.

Q. (Jackson) In that case, how do you locate the subset in the locality of which you search for an optimum?

A. We obtain local optimum by an algorithm of transfer, which moves one object at a time until the local optimum is reached.

Comment: (Roux) The algorithm about which M. Lerman spoke begins by dividing the set to be classified into ten or more arbitrarily classes and then taking one element and trying to put it in the better class, *i.e.* that which maximises the function given by M. Lerman.

Q. (Cole) Do you use the concept of linear ordination only in the proof of your theorem or also in the practical algorithm? If you use it in the algorithm does not this destroy the concept of taxonomic distance?

A. Yes, but it's of practical interest. I think we should take the pre-order on F rather than a distance which in taxonomy is always arbitrary.

Q. (Saksena)

1. Could you illustrate with examples that your approach gives better classification than the existing methods of numerical taxonomy?
2. From what you have stated, it appears, that you even doubt the validity of the idea of taxonomic distance as a means of classification; am I right in concluding this?

A.

1. Yes, I have several and I can send you the details.
2. I think that distance is a topological notion. If we can do without using distance, so much the better. In any case I need some parameter to establish my pre-order. I can show that my pre-order is more stable.

A TAXIMETRIC APPROACH TO THE CLASSIFICATION OF THE SPINY-FRUITED MEMBERS (TRIBE CAUCALIDEAE) OF THE FLOWERING-PLANT FAMILY UMBELLIFERAE

J. McNeill, P. F. Parker and V. H. Heywood

The Hartley Botanical Laboratories, University of Liverpool
Department of Botany, University of Leicester
Department of Botany, University of Reading

The work which is described in this paper is a small part of much more extensive research undertaken at Liverpool on the principles and practice of taxonomy, utilising members of the angiosperm family Umbelliferae (the carrot family). This work has several facets; some of it, for example, is directed towards assessing the value of chemical data in taxonomy (cf. Harborne, 1967a, 1967b), while some is exploring whether a genuinely phenetic classification, utilising all possible characters without *á priori* weighting, is compatible with the traditional restriction to visual characters, for reasons of expediency, in the practical classification of the flowering plants. It is with this facet that the present studies are associated.

Silvestri and Hill (1964) list three desiderata of a biological classification: it should have *stability*, being little affected by new data, *objectivity*, similar results being obtained by taxonomists working independently on the same organisms, and high *predictivity*. By this last requirement is meant that a good general biological classification should enable one to predict with a high degree of accuracy the distribution within the group of attributes later found in some members not originally used in the construction of the classification (Heywood, 1967).

The classification of the flowering plant family Umbelliferae at generic level and above is traditionally based mainly on characters of the fruit and to a lesser extent on the details of inflorescence structure. Does the distribution of characters from other parts of the plant show a reasonable conformity with the existing classifications and what would be the result of constructing a classification from a range of characters in which those of the fruit were excluded altogether? These are among the questions which we have been asking in our work in Liverpool.

Existing Classification of the Caucalideae

The part of the Umbelliferae with which we have been particularly concerned comprises those spiny-fruited species, with both primary and secondary ridges on the fruit, which Bentham in Bentham and Hooker in their *Genera Plantarum* (1867) and those who followed them such as Boissier (1872) regarded as forming the tribe Caucalideae. They include the carrot itself *(Daucus carota)* and about 75 other species usually arranged today in about 18 genera, all extremely small except for *Torilis* (15-20 species, including Hedge-parsley in Britain) and *Daucus* (c.25 species). The group is centred in Europe, North Africa and South-West Asia, and the fact that this allows a wide range of living material to be obtained was an important factor in its selection. In addition to this and to its convenient size (*i.e.* a relatively low number of species with very considerable generic diversity) the other important reason for selecting this group is that it has been very differently classified by other authors, notably by Drude (1898) in the other standard systematic treatise extending to generic level, Engler and Prantl's *Die natürlichen Pflanzenfamilien*.

Drude recognised two groups where Bentham and Hooker and Boissier had only one, and he went so far as to place these at almost opposite ends of his generic arrangement of the family. One group, his tribe Dauceae, consisting of *Daucus* and four other small genera, all with spines on the primary and particularly on the secondary ridges of the fruit, he believed to have evolved from plants such as those included in the tribe Laserpiteae, whose members have fruits without spines, but with primary and prominent secondary ridges. The other group he linked closely with genera such as *Anthriscus* and *Scandix* (Bentham's tribe Scandiceae) which lack secondary ridges. Drude regarded the two groups as subtribes (Scandicinae and Caucalinae) of a single tribe (Scandiceae) and believed that the secondary spinose ridges in his Caucalinae had evolved independently from those in his Dauceae. The dividing line in fruit morphology is far from sharp and Drude recognises in his Caucalinae two segregate genera (*Orlaya* and *Astrodaucus*), which Bentham treated as constituents of the genus *Daucus*.

Although there has been no monograph or full revision of the Umbelliferae or of this part of the family since Drude (1898), other authors such as Calestani (1905) and Cerceau-Larrival (1962, 1965) have proposed partial arrangements of the family, which do not entirely match either Bentham's or Drude's treatment (Table 2).

In addition to this classificatory conflict at tribal and sub-tribal level there has been considerable dispute as to generic limits, in particular in determining the boundary between *Caucalis* and *Torilis* and in the total number of recognisable genera, the numbers adopted varying between eight and twenty. It is against this background that the taximetric work has been undertaken. We have not concerned ourselves with taxonomic problems at the specific or infra-specific levels, though these do exist (notably in the *Daucus carota* and *Torilis arvensis* groups) and we will in effect treat species or sometimes subspecies as our operational taxonomic units (OTUs), although in the work reported here we have in fact used the plants raised from individual seed accessions as our OTUs representing the species or subspecies.

Choice of Taximetric Method

An enumeration was made of the characters which had previously been used in the classification of the Caucalideae at any level and to this was added other attributes, many of them from the leaves, which were seen to vary between species. While many of these characters chosen, such as the presence or absence of a particular flavone, were qualitative (*i.e.* two-state or *binary*) the majority were quantitative (numerical), being linear measurements or numbers of parts; there was also a considerable number of multi-state characters, such as the type of hair present on the stem, in which the different states have no evident quantifiable linear relationship with one another.

At the time when this work was started, the MULTIST program developed at Canberra by G. N. Lance and W. T. Williams (cf. Ducker, Williams and Lance, 1965; Lance and Williams 1966a) was one of the few which avowedly sought to deal with data of these three types, such as would normally arise in any taximetric study involving a group of morphologically elaborate organisms such as the angiosperms. This program utilised the 'non-metric coefficient' $(b+c)/(2a+b+c)$, (cf. Lance and Williams 1966b) and the 'centroid' sorting strategy (cf. Lance and Williams, 1966a).

The program and possible developments from it, some of which are now operational and reported here, seemed well-suited to our purpose and we were fortunate in obtaining the generous co-operation of the Computing Research Section of the C.S.I.R.O., Canberra in agreeing to process our data.

OTUs and Characters

Scoring was begun for most of the characters noted (the only notable exception being the pollen characters used by Cercêau-Larrival, 1962, 1965) on the plants grown from seed accessions at the University of Liverpool Botanic Garden, Ness. A small sample of these, 27 accessions, (cf. table 1) was chosen for a trial computer analysis to assess whether any modification or extension of the characters scored was necessary and whether the program grouped different accessions of the same species satisfactorily.

The sample of 27 accessions represents 13 species referable to eight genera, all except two of the accessions being generally regarded as members of the Caucalideae (or Dauceae + Caucalinae in Drude's terminology) (cf. Table 2). The two exceptions (Accession nos. 12 and 25) are *Physocaulis nodosus*, a member of the Scandiceae *sensu stricto*, included as a 'marker' member of that group and *Coriandrum sativum* (Coriander) a spineless species which Bentham and Hooker (1867) included in the Caucalideae but which later authors including Drude (1898) and Calestani (1905) have placed with a few other genera in a separate tribe, the Coriandreae.

Analysis No.	Species	Accession No.
1.	Chaetosciadium trichospermum (L.) Boiss	2945
2.	Daucus carota L.	1691
3.	" "	1750
4.	" "	1780
5.	" "	1858
6.	" "	2045
7.	" muricatus (L.) L.	1782
8.	" "	2458
9.	" littoralis Sibth. & Sm.	2948
10.	" glochidiatus (Labill.) Fischer & Meyer	1771A
11.	" durieua Lange	2889
12.	Physocaulis nodosus (L.) Tausch	1875
13.	Orlaya grandiflora (L.) Hoffm.	1883
14.	" "	1747
15.	Daucus muricatus (L.) L.	2861
16.	Orlaya grandiflora (L.) Hoffm.	1763
17.	Torilis arvensis (Hudson) Link	1749
18.	" "	1784
19.	" "	1817
20.	" "	2085
21.	" nodosa (L.) Gaertner	2084
22.	Turgenia latifolia (L.) Hoffm.	2205
23.	" "	2334
24.	" "	2016
25.	Coriandrum sativum L.	3221
26.	Caucalis platycarpos L.	1827
27.	" "	1829

Table 1. O.T.U.'s analysed

Table 2 Taxonomic treatments of species used in analysis

Genera in analysis	Total No. of species	No. of species in analysis	No. of OTUs in analysis	Bentham & Hooker (1867)		Drude (1898)		Calestani (1905)	
DAUCUS L.				Daucus	CAUCALIDEAE ('Caucalineae')	Daucus	DAUCEAE	Daucus	DAUCEAE
Sub-genus Durieua	7–8	2	2						
Sub-genus Daucus	c.15	3	9						
(sub-genus Ctenolophus)	2–3	0	0						
ORLAYA Hoffm.	3–4	1	3			Orlaya	SCANDICEAE: CAUCALINAE	Orlaya	
TORILIS Adanson	15–20	2	5	Torilis		Torilis		Torilis	CAUCAL-IDEAE
TURGENIA Hoffm.	1	1	3	Caucalis		Caucalis		Turgenia	
CAUCALIS L.	2–3	1	2					Caucalis	
CHAETOSCIADIUM Boiss.	1	1	1			Chaetosciadium		Chaetosciadium-CHAETOSCIADEAE	
CORIANDRUM L.	1	1	1	Coriandrum		Coriandrum-CORIANDREAE		Coriandrum-CORIANDREAE	
PHYSOCAULIS (DC.) Tausch	1	1	1	Physocaulis-SCANDICEAE		Physocaulis-SCANDICEAE: SCANDICINAE		Physocalis-CAUCALIDEAE	

Table 2. Taxonomic treatments of species used in analysis

Of the other accessions, no.1, *Chaetosciadium trichospermum* has most striking bristly fruits and although the species was included within *Caucalis* by Bentham and Hooker, Calestani has placed the monotypic genus *Chaetosciadium* in a sub-tribe of its own, the *Chaetosciadeae*. Four and five accessions, respectively, of the variable *Torilis arvensis* and *Daucus carota* are included and within the genus Daucus two of the other accessions (nos. 10 and 11) represent members of the sub-genus *Durieua*. Both of the Torilis species represented are members of the same section (*Lappularia*).

83 characters were used in the analysis and although a high proportion did come from fruit morphology there were also large groups of leaf and inflorescence characters. The numbers from each part of the plant are listed in table 3 as are their contribution to binary, numerical and multistate characters.

	Binary	Quantitative	Multistate	Total	%
Indumentum	0	3	1	4	5
Stem	1	1	0	2	2
Leaf	0	18	1	19	23
Inflorescence	2	13	0	15	18
Flower	1	6	1	8	10
Fruit	3	17	11	31	37
Chemistry	4	0	0	4	5
Total	11	58	14	83	100

Table 3. Sources of Characters used in Analysis

Computer Programs

The data for this pilot analysis were processed in June 1967, by which time the Canberra MULTIST program, already referred to, had been replaced by the more versatile program MULTCLAS. This is one of the programs used by El-Gazzar, Watson, Williams and Lance (1968) and is essentially the program MULTIST retaining the non-metric coefficient as the similarity measure but modifying the sorting strategy from a centroid one to a 'flexible' one (cf. Lance and Williams (1967)). In the system erected by Lance and Williams (1966c) the distance between an existing group h and a group k, formed by the fusion of i and j is of the form:

$$d_{hk} = \alpha_i d_{hi} + \alpha_j d_{hj} + \beta d_{ij} + \gamma |d_{hi} - d_{hj}|$$

All the established linking strategies such as nearest-neighbour, median group-average and centroid can be expressed in this form with varying values of α_i, α_j, β and γ. The 'flexible' sorting strategy used in MULTICLAS operates within this system with the quadruple constraint that

$$\alpha_i + \alpha_j + \beta = 1; \ \alpha_i = \alpha_j; \ \beta < 1 \ ; \ \gamma = 0$$

In the sorting of our data the parameter values used were $\alpha_i = 0.625$; $\alpha_j = 0.625$; $\beta = -0.25$. A small negative value of β appears in general to give the most space conserving sorting. Certainly when β approached 1 'chaining', in which OTUs link on singly to the group first formed, seems complete (cf. Lance and Williams, 1967a, fig. 1).

MULTBET, the other program by which the data were analysed is a more recent development which utilises one of three alternative similarity measures, the Canberra metric, the Gower metric, or a Shannon-type information statistic (cf. Lance and Williams, 1967b). In the analysis of our data the information statistic was used, the continuous numerical data being divided into eight states for this purpose. Centroid sorting alone is used, the other alternatives available in MULTCLAS being incompatible with the information statistic.

Results of Computer Analyses - 1) Dendrograms

As is evident from figs. 1 and 2 the correspondence between the dendrograms produced by the two programs is fairly close and moreover both are reasonably in accord with those parts of the taxonomy of the group which are not in doubt.

Most of the accessions referable to a single species are sorted out together, the exceptions being the intermingling of *Chaetosciadium* in both programs, and *Torilis nodosa* in MULTBET, with *Torilis arvensis* and the separation in MULTCLAS of one rather odd population of *Daucus carota* from the others referable to this species.

The most striking features of the overall pattern however are a) the isolation of *Turgenia* in MULTCLAS, b) the fairly close association of *Physocaulis* and *Coriandrum* with some of the species invariably included in the Caucalideae and c) the suggestion of a major separation between a group consisting of *Orlaya* and *Daucus* subgenus *Daucus* on the one hand and all the other taxa, including *Torilis*, *Caucalis* and *Daucus* subgenus *Durieua* on the other.

The isolation of *Turgenia* might at first sight seem surprising as the one species (*T. latifolia*) has often been included in the genus Caucalis. On the other hand it has broad cotyledons (cf. Cercêau-Larrival, 1962), an unusual microstructure as revealed by scanning electron microscopy (Heywood, unpublished) and a distinctive chromosome base number ($n = 16$), which are not known elsewhere in the group and which were not characters incorporated in the analysis.

The failure to separate *Physocaulis* and to a lesser extent *Coriandrum* probably shows up the type of weakness in our method for which this preliminary analysis was designed. The great majority of characters used are those that vary within the Caucalideae and although the more obvious unique characters of Physocaulis were included it is likely that some others have been overlooked. It is possible also that the size of the groupings had an effect on the sorting strategy. There were 25 OTU's drawn from the Caucalideae as against one *Physocaulis* and one *Coriandrum*.

If this reflected the total numbers of species to be classified is it not possible that taxonomists would also lump these together in a single group?

The suggestion of a major dichotomy within the sample is of course the most interesting result particularly as the division would fall in a very different position from that postulated by Drude (1898), though somewhat nearer that suggested by Calestani (1905). It is clear, either that there is a hitherto unsuspected discontinuity here which the taxonomist must explore, or else that this is an artifact of the smallness of the sample. In either case Drude's conclusions seem suspect while judgement must be reserved on Bentham's.

The close association of *Chaetosciadium* with *Torilis* in the output of both programs suggests that Calestani's recognition of a tribe Chaetosciadeae is scarcely warranted. A detailed study of the genus from the point of view of anatomy and microstructure (Heywood, unpublished) and chemical constituents (Crowden *et al*, 1969) has confirmed its similarity to *Torilis*.

Results of Computer Analysis - 2) Grouper Analysis

In addition to a dendrogram plot the output of both MULTCLAS and MULTBET includes a comparison of the character distributions in each of the major fusing groups. This 'grouper' analysis expresses the contribution of each character in discriminating two groups which fuse, as the absolute difference between the character's proportionate frequence (on a 0 to 1 scale in the two groups). Thus a character consistently present in one group and absent in the other would have a value of 1 and one that was equally frequent in the two groups a value of 0.

This analysis thus provides a measure of the contribution of each character to the classification. In this study, the output, in fact, provides a grouper analysis of the first three dichotomies of the MULTCLAS dendrogram, and the first four of that of MULTBET. This allows us to attempt an answer to the question of whether fruit characters really do demarcate primary groups or whether they represent an unsubstantiated taxonomic tradition.

	All characters used	Characters with difference ≥ 0.5	Top Twenty	Top Ten
Indumentum	5	15	10	10
Stem	2	0	0	0
Leaf	23	0	0	0
Inflorescence	18	23	25	20
Flower	10	15	30	20
Fruit	37	46	35	50
Chemistry	5	0	0	0

Table 4. Sources of Characters (%) Contributing Most to First Dichotomy (MULTBET)

	All characters used	Characters with difference ≥ 0.5	Top Twenty	Top Ten
Indumentum	5	11	9	10
Stem	2	0	0	0
Leaf	23	3	3	0
Inflorescence	18	20	19	25
Flower	10	8	18	8
Fruit	37	54	48	43
Chemistry	5	6	5	5

Table 5. Sources of characters (%) contributing most to first 4 dichotomies (MULTBET)

Except where groups comprise only a single genus, such as the *Turgenia* group, a character value of 1 is rare but Tables 4 and 5 show the parts of the plant from which were derived the characters which contributed most to the first and the four primary MULTBET dichotomies respectively. The situation with the MULTCLAS dichotomies is similar.

It is at once evident that fruit characters have contributed more to primary group delimitation than would be expected purely by chance. On the other hand leaf characters which made up about 20% of the total have contributed virtually nothing to primary group delimitation.

It might be suggested that the original large proportion of fruit characters has biassed the analysis towards a classification based principally on fruit morphology, but this scarcely matches the unexpected relative importance of floral characters, which made up only a small proportion of the total.

The most reasonable explanation of the results of the grouper analysis, is that the traditional taxonomic reliance on fruit and inflorescence characters in this group is well-founded that leaf characters are of little value, while the minute flowers can contribute much more valuable comparative data than has hitherto been realised.

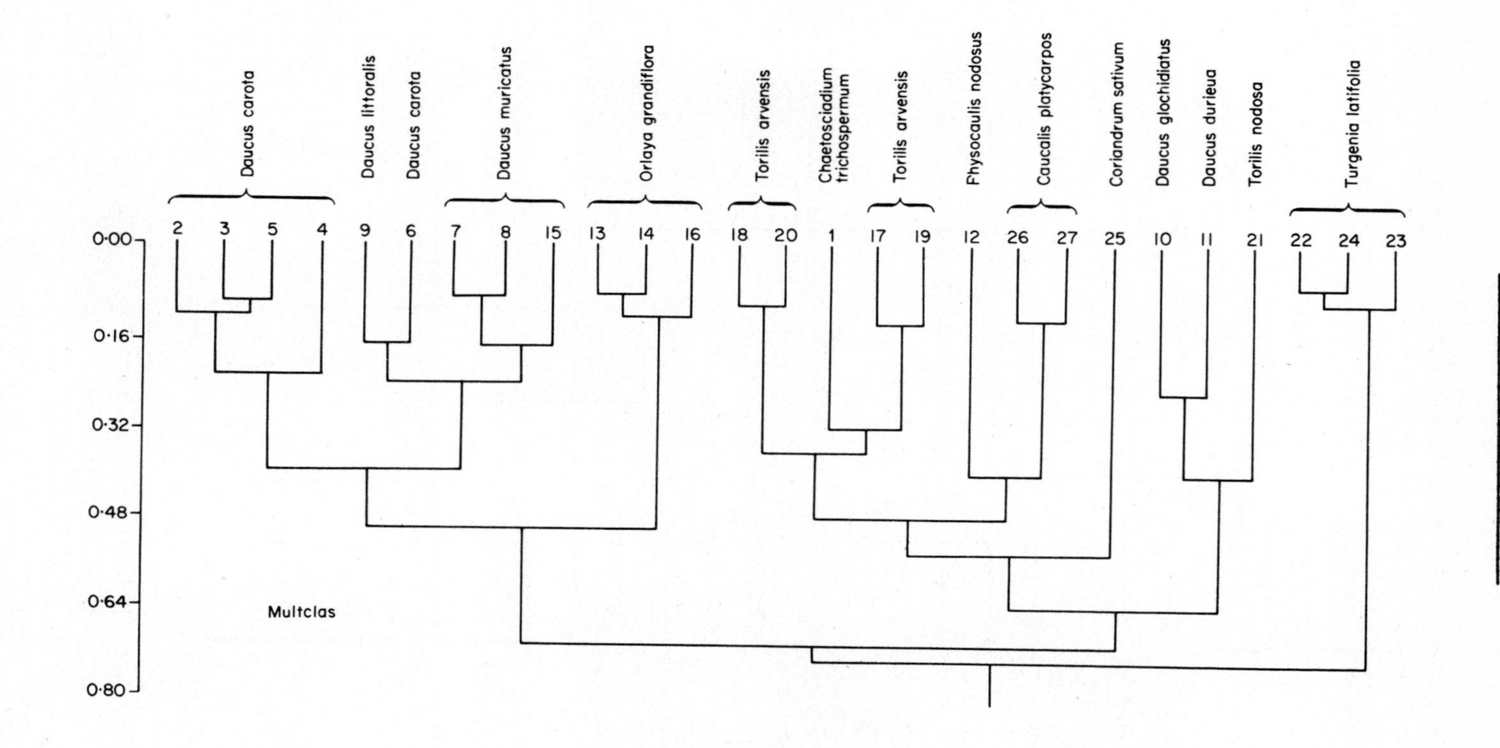

Fig. 1. MULTCLAS

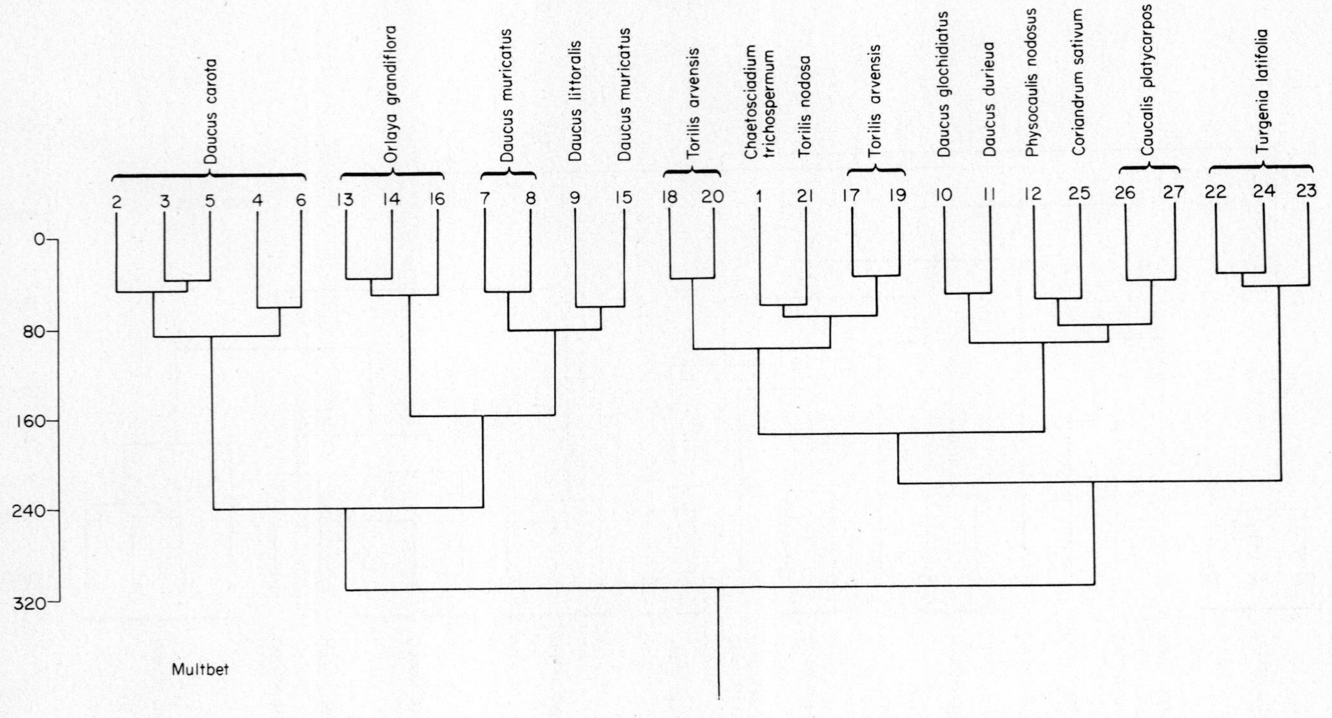

Fig. 2. MULTBET

Conclusions

This preliminary analysis, while revealing several small defects in character selection and scoring, has shown that in general the Canberra mixed-data programs provide a reasonably satisfactory treatment of the sample analysed. A second analysis incorporating a further nineteen O.T.U.'s and covering a wider range of genera and species is in progress at the moment and indeed we had hoped to be able to report on this here. This second analysis covering 18 species and 10 genera should throw much more light on the taximetric relationships of the members of the Caucalideae.

We do not feel competent to assess whether the non-metric coefficient with its original rejection of negative matches (though this does not seem to apply to the MULTIST version given by Lance and Williams, 1966a) is the most appropriate for taxonomic purposes, nor whether the apparently pragmatic sorting strategy used in this example approximates at all closely to the way in which a taxonomist believes group formation to be attempted. As we have indicated our choice of the Canberra programs was largely determined by capability of handling mixed data and their availability. We agree with Williams and Lance (1965) when they point out that "taxonomists cannot be expected to use numerical methods until these are capable of processing their data without distortions imposed solely for the convenience of the method". Equally as taxonomists we must enunciate more clearly our requirements in group formation and particularly in the balance that we want to achieve between parameters such as group size and dispersion as well as inter-group distance.

Acknowledgement

The work upon which this contribution is based was undertaken between 1965 and 1968 as part of research on a multivariate Approach to Angiosperm Classification supported by the United Kingdom Science Research Council research grant B/SR/1923 to Professor V. H. Heywood, Dr. J. McNeill and the late Dr. H.M. Hurst.

This research is continuing at the University of Reading under an extension of the grant.

References

Bentham, G. and Hooker, J.D., 1867, *Umbelliferae, in Genera Plantarium 1* : 859-831. London; Reeve.

Boissier, E., 1872, *Umbelliferae, in Flora Orientalis 2* : 819-1090 Genevae and Basilae: H. Georg.

Calestani, V., 1905, Contributo alla sistematica dell Ombellifere d'Europa. *Webbia 1* : 89-280.

Cercêau-Larrival, M.T., 1962, Plantules et pollens d'Ombelliferes. *Mem. Mus. natn. Hist.* nat., Paris, ser. B (Bot) *14* : 1-166.

Cercêau-Larrival, M.T., 1965, Le pollen d'Ombelliferes Mediterraneennes III - Scandicineae Drude, IV. - Dauceae Drude. *Pollen & Spores 7* : 35-62.

Crowden, R.K., Harborne, J.B., and Heywood, V.H., 1969, Chemosystematics of the Umbelliferae, I. A General Survey. *Phytochemistry* (in Press).

Drude, O., 1898, Umbelliferae: Apioideae, in A. Engler and K. Prantl, *Die natürlichen Pflanzenfamilien 3 (8)*: 145-250.

Ducker, S.C., Williams, W.T. and Lance, G.N., 1965, Numerical classification of the Pacific forms of *Chlorodesmis* (Chlorophyta). *Aust. J. Bot. 13* : 489-499.

El-Gazzar, A., Watson, L., Williams, W.T., and Lance, G.N., 1968, The taxonomy of *Salvia* : a test of two radically different numerical methods. *J. Linn. Soc. (Bot) 60* : 237-250.

Harborne, J.B., 1967a, Comparative biochemistry of flavonoids V Luteolin 5-glucoside and its occurrence in the Umbelliferae. *Phytochem. 6* : 1569-1573.

Harborne, J.B., 1967b, Flavonoids of the Umbelliferae, in *Comparative Biochemistry of the flavonoids* pp. 180-183. London & New York: Academic Press.

Heywood, V.H., 1967, *Plant Taxonomy*. London: Edward Arnold.

Lance, G.N. and Williams, W.T., 1966a, Computer programs for classification. *Proc. 3rd Aust. Comp. Conf. Canberra* Paper 12/3 3 pp.

Lance, G.N. and Williams, W.T., 1966b, Computer programs for hierarchical polythetic classification ("similarity analyses"). *Computer J. 9* : 60-64.

Lance, G.N. and Williams, W.T., 1966c, A generalised sorting strategy for computer classifications. *Nature, Lond. 212:* 218.

Lance, G.N. and Williams, W.T., 1967a, A general theory of classificatory sorting strategies. 1. Hierarchical systems. *Computer J. 9* : 373-380.

Lance, G.N. and Williams, W.T., 1967b, Mixed-data classificatory programs. I. Agglomerative systems. *Aust. Computer J. 1* : 15-20.

Silvestri, L. and Hill, I.R., 1964, Some problems of the taxometric approach, in V.H. Heywood and J. McNeill (eds). Phenetic and phylogenetic classification pp. 87-104. London : Systematics Association. *(Syst. Assoc. Publs. 6)*.

Walters, S.M., 1961, The shaping of angiosperm taxonomy. *New Phytol. 60* : 74-84.

Williams, W.T. and Lance, G.N., 1965, Logic of computer-based intrinsic classifications. *Nature, Lond. 207* : 159-161.

Discussion

Comment: Barrs confirmed that the MULTIST version at Canberra used a non-metric coefficient modified (in some way) to take account of double negative matches.

Q.(Wishart) If you have used other options of the Lance-Williams program could you say how you arrived at the particular dendrograms you have shown us, since the groupings of the individual methods vary widely? In addition the flexible form also produces different groupings dependent on the value of β which one chooses.

A. In fact we haven't used the other options as yet, we have been dependent on the Canberra people as to what is run. We want now to do a series of runs of this sort on our batch of data. You are correct that changing the value of β results in complete chaining to beautiful dichotomies as β varies between values approaching 1 down to - 0.1 I would refer you to the set of dendrograms in Lance and Williams, (1967a) which demonstrate this.

Q. (Hall) In experience with the information statistic we have found a marked tendency to form groups symmetrical in terms of numbers of OTU's. Are there indications of this in your work?

A. Yes, but I have insufficient knowledge of the mathematics of the information statistic to know if this is a property of it, but I agree that in looking at the results from a purely taxonomic view it is incredibly symmetrical. The species included in the analysis were those which were ready at the time, and not a representative sample in the taxonomic sense. We would, therefore, expect a more untidy result in reality.

Comment: (Barrs) At Southampton we have an information analysis program and this tendency has been noted. However, I do not know if this is inherent in the information statistic or whether it results from the Centroid sorting procedure, as I suspect. In larger sets of data the phenomenon is marked at a lower level of the hierarchy but as one proceeds towards a single group it becomes less noticeable.

Q.(Crawford) Have you tested the effect of adding new data to your results - for instance have you considered analysing the British umbellifers and then adding the rest of them? If you had to criticize your results 'in vacuo' how do you think they would stand the addition of more data?

A. Undoubtedly they would change. However, I consider that the criterion of stability to which I referred, was defined in terms of adding new attributes rather than of adding new individuals. We think that addition of new attributes would affect the structuring very little, but addition of more than a very few OTU's must have a marked effect. Any taxonomic treatment is affected by the size of the group; one slightly anomalous individual OTU's would be incorporated in an existing group but fifty all differing slightly from each other though at a similar distance from an existing group should be recognised as a separate group; this applies equally to a numerical approach or a traditional taxonomic one. There is, in fact, a suggestion (Walters, 1961) that if the classification of the Umbelliferae had originated not in N. Europe in the 18th Century (where they are a fairly dominant family and the Araliaceae are barely represented) but say in N Zealand in the 19th century then they would merely be one genus of the Araliaceae.

INFORMATION THEORY MODELS FOR HIERARCHIC AND NON-HIERARCHIC CLASSIFICATIONS

Laszlo Orloci

Department of Botany, University of Western Ontario

Abstract

Several information theory functions are described that are suitable for the imposition of structure upon multivariate collections. These functions include total information, joint information and mutual information. Furthermore, the application of discrimination information is discussed in connection with non-hierarchic sorting of individuals between classes of known parameters. And regarding hierarchic classifications, the use of heterogeneity information, as a decision function in subdivisive and agglomerative clustering, is outlined.

Introduction

It seems appropriate to begin this presentation with some general remarks regarding the concept of information and information analysis. Firstly, it must be made absolutely clear that the term information is understood here in a strictly technical sense. It is in fact conceived as a physical property of events related to probability. Note, that in accordance with this definition a rare event conveys more information than a common event. Information as a technical term thus relates conceptually more closely to surprisal than to either knowledge or informativeness of the ordinary speech.

The conventional techniques of classification frequently operate within the framework of different restrictive assumptions. These assumptions often relate to the type of the data to be analysed, the nature of the underlying probability distributions, the size, shape and orientation of the hyperclusters in sample space, or the properties of the sample space itself. Consider, for example, discriminant analysis as presented by Fisher, (1936) or Rao (1952). This technique requires the assumption that (1) a classification exists and the problem only amounts to sorting individuals between the existing classes according to an arbitrary definition of best fit, (2) the classes correspond to populations each with an underlying (multivariate) normal distribution, (3) the populations, geometrically represented as hyperclusters, only differ from each other by the spatial orientation and length of their mean vectors, but they are otherwise equivalent in every respect including shape, size and orientation, and (4) the sample space is strictly Euclidean. Restrictive assumptions also exist, but perhaps to a lesser extent, in the techniques of cluster analysis. The use of a metric definition of structure (*e.g.* Macnaughton-Smith et al. 1964; Ducker et al. 1965; Edwards and Cavalli-Sforza 1965; Jancey 1966; Orloci 1967), or restriction of the analysis to specific types of data (*e.g.* Williams and Lambert 1959) are two examples in this connection. It is understood that these assumptions, particularly (3), may restrict the application of a classification technique to the typical case that is rarely found in natural situations. It is nevertheless recognized that valid alternative procedures can be found that may effectively replace the classical techniques. The alternative procedures, that are founded in information theory, are particularly noteworthy, regarding taxonomic applications because of their extreme flexibility in being easily applied to practical problems with no assumptions required about the data, the underlying probability distribution, the size, shape or orientation of the hyperclusters, or the properties of the sample space at hand. It is exactly for these reasons that the techniques of information theory are so much favoured by an increasing number of taxonomists in the different fields.

Definitions of Structure

General concepts

For reasons of convenience, a taxonomic collection is conceived as a set of individual frequency distributions, each representing an individual, or a character, depending on the problem at hand. Information theory offers several distinct functions suitable for the definition of structure in such a collection. These functions include total information, joint information and mutual information, which are related according to Abramson's (1963) diagram given in Fig. 1. In this diagram $I(h)$ and $I(i)$ indicate the total information within the frequency distributions labelled X_h and X_i respectively.

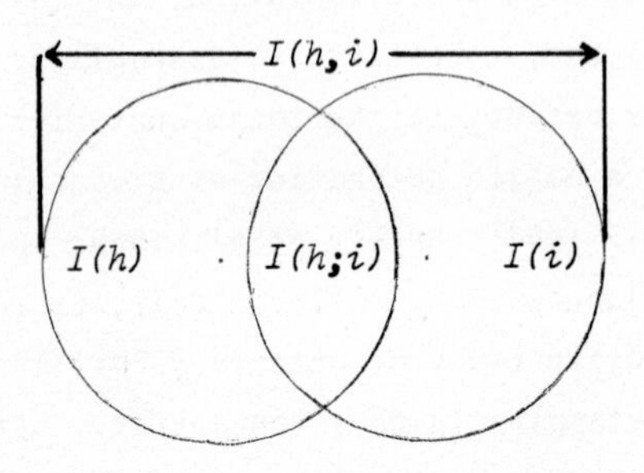

Fig. 1. Diagramatic representation of information. After Abramson (1963).

$I(h,i)$ and $I(h;i)$ on the other hand, represent the joint information and mutual information between X_h and X_i. It will be assumed that the collection consists of r frequency distributions. These distributions are independent, or related, as the case may be. Each distribution is conceived as a consequence of the union of several subsets of frequency classes, X_{hA}, X_{hB}, and so forth. The union may take place in a manner such that the frequencies are pooled between the classes that possess equal class values or identical class symbols. In specific situations, however, such pooling of frequencies is not justified if the subsets are qualitatively disjoint.

Total information in X_h

A much used measure for total information is Shannon's (1948) entropy function. This function can be written for the hth frequency distribution X_h in the form

$$I(h) = - \sum_{j=1}^{n_h} f_{hj} \ln p_{hj} = N \ln N - \Sigma f_{hj} \ln f_{hj} \qquad (1)$$

where the *f*s are class frequencies, of which there are n_h, the *p*s are independent *á posteriori* probabilities, and N is the total number of observations such that

$$N = \Sigma f_{hj} \quad \text{and} \quad p_{hj} = f_{hj}/N .$$

If X_{hA} and X_{hB} are subsets of classes such that the union of X_{hA} and X_{hB} is equal to X_{hAB} then the following relation is of importance:

$$I(h)_{AB} \geq I(h)_A + I(h)_B . \qquad (2)$$

The quantity on the left hand side, representing the total information conveyed jointly by the two subsets in h, is never less than the pooled information conveyed separately by the subsets. This relation has been used by different authors (*e.g.* Macnaughton-Smith 1965; Williams et al. 1966; Orloci 1968) to find optimal unions, or subdivisions, in biological collections.

Joint information between frequency distributions

In some applications the r *I*s are pooled to derive an overall measure for joint information:

$$I(1,2,\ldots,r) = \sum_{h=1}^{r} I(h) \qquad (3)$$

Note, however, that the pooling of the individual terms of information is only valid if the r distributions are independent. In the alternative case, the pooled value of information exceeds the actual joint information with an amount equal to the mutual information shared between the r distributions.

It is always possible to represent any r frequency distributions, and the relationship between them, in an r

dimensional contingency table. In such a table the original class frequencies appear as r sets of marginal totals, and the values inside the body of the table specify the frequencies of the joint observations made on the r entities simultaneously. When $r = 2$, the common case of paired comparisons, then the information conveyed jointly by X_h and X_i is given by

$$I(h,i) = - \sum_{j=1}^{n_h} \sum_{k=1}^{n_i} f_{jk} \ln p_{jk} = N \ln N - \Sigma\Sigma f_{jk} \ln f_{jk} . \quad (4)$$

A similar expression can be found for joint information $I(1,2,\ldots,r)$ in the case of any r dimensional table. The symbols used in equation 4 correspond to those given in Table 1.

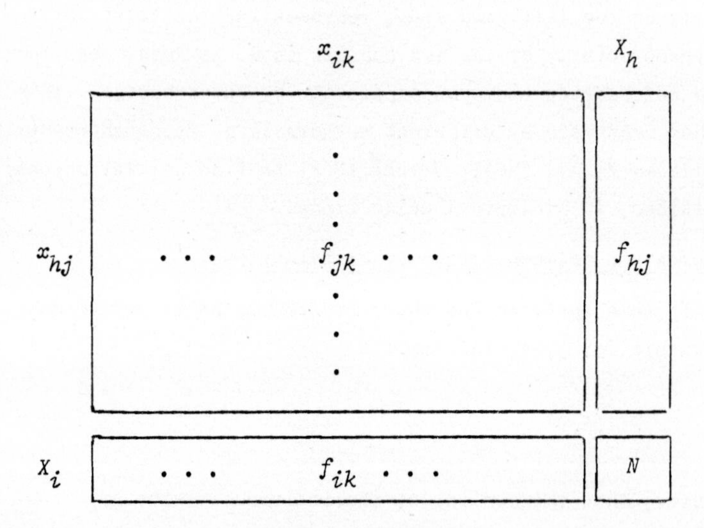

Table 1. Contingency table relating two frequency distributions X_h and X_i

The *á posteriori* probabilities are defined by $p_{jk} = f_{jk}/N$. The sum N indicates the total number of paired-values realized as actual observations between the two frequency distributions.

The f_{jk} is short for f_{hijk}.

If the contingency table is visualized as that containing the relationship between subsets X_{hA} and X_{iA}, and if the relationship between subsets X_{hB} and X_{iB} is represented by a second table, then the following relations apply:

$$I(h,i)_{AB} \geq I(h,i)_A + I(h,i)_B \qquad (5)$$

$$I(h,i)_{AB} \leq I(h)_{AB} + I(i)_{AB}$$

or

$$I(h,i)_A \leq I(h)_A + I(i)_A \; . \qquad (6)$$

The quantity $I(h,i)_A$ corresponds to a contingency table relating X_{hA} and X_{iA}. Similarly $I(h,i)_B$ corresponds to a table relating X_{hB} and X_{iB}. In this context, $I(h,i)_{AB}$ corresponds to a table that can be derived by the union of the two tables corresponding to the subsets with or without pooling frequencies between them, depending on the nature of the problem at hand.

The inequality given as expression 5 plays an important role in selecting optimal fusions or subdivisions in cluster analysis, considered in some detail later in this presentation. Expression 5 actually implies that the value of the joint information in the united subsets $A + B$ cannot be less than a quantity obtained by pooling the joint information corresponding to the subsets. The inequality given as expression 6, a consequence of the relations shown in Fig. 1, is discussed in detail by Khinchin (1957). Note that the joint information, given as equation 4, actually differs from the pooled total information (equation 3) by a correction for mutual information..

Mutual information

The information possessed in common between two frequency distributions X_h and X_i, called mutual information, can be expressed in terms of the symbols given in Fig.1. The corresponding algebraic expression simplifies to

$$I(h;i) = \sum_{j=1}^{n_h} \sum_{k=1}^{n_i} f_{jk} \ln \frac{N f_{jk}}{f_{hj} f_{ik}}$$

$$= \Sigma\Sigma f_{jk} \ln f_{jk} + N \ln N - \sum_{j=1}^{n_h} f_{hj} \ln f_{hj} - \sum_{k=1}^{n_i} f_{ik} \ln f_{ik}. \qquad (7)$$

Eq. 7 is known as the error or independence component of the discrimination information (Kulback and Leibler 1951; Kulback 1959; Kulback et al 1962). In the case of an r dimensional contingency table the simplest expression for the overall mutual information is

$$I(1;2;...;r) = \sum_{h=1}^{r} I(h) - I(1,2,...,r).$$

When N is sufficiently large then the mutual information is asymptotically distributed χ^2 at $(n_h - 1)(n_i - 1)$ degrees of freedom for a two dimensional table, or $n_1 n_2 ... n_r - n_1 - n_2 ... - n_r + r - 1$ degrees of freedom in the case of an r dimensional contingency table.

The following relations are of importance:

$$I(h;i)_{AB} \geq I(h;i)_A + I(h;i)_B \qquad (8)$$

and

$$I(h;i) \leq I(q)$$

where $I(q)$ is the smallest of $I(h)$ and $I(i)$. Similar expressions can be easily derived also for the r dimensional case. Relation 8 actually implies that the mutual information in the united subsets $A + B$ cannot be less than a quantity obtained by pooling the mutual information corresponding to the subsets. The inequality in expression 9 is a direct consequence of the relations in Fig.1.

The relative relatedness of two frequency distributions can be measured by Rajski's coherence coefficient (Rajski 1961) which is given for X_h and X_i by

$$R(h;i) = (1-d^2(h;i))^{\frac{1}{2}} \tag{10}$$

where $d(h;i) = 1 - \frac{I(h;i)}{I(h,i)}$ called Rajski's metric. The value of $R(h;i)$ varies between zero and unity, indicating respectively the degree of relatedness from none to perfect. The probability corresponding to $2I(h;i)$ may also be mentioned as a relative measure of relatedness.

Example

As an example consider the data in Table 2 indicating the performance of 2 species in 28 stands of vegetation evaluated in accordance with an arbitrary scale. The contingency table relating the two species of Table 2 is given in Table 3. The column totals in this table represent the observed frequency

Species	Symbols	Subset *A*												
		1	2	3	4	5	6	7	8	9	10	11	12	13
Artmisia maritima L.	*h*	3	3	3	3	3	3	3	0	+	0	0	0	3
Kraschenìnnikovia ceratoides Güld.	*i*	1	+	1	+	+	+	1	2	2	2	2	2	1

Species	Symbols	Subset B														
		14	15	16	17	18	19	20	21	22	23	24	25	26	27	28
Artemisia maritima L.	h	3	3	3	3	3	2	4	3	3	3	3	3	+	4	2
Krascheninnikovia ceratoides Güld.	i	+	+	+	+	1	0	1	1	2	0	0	0	0	+	0

Table 2. Species performance data in 28 stands.*

*Data after Hartman (1968) Scale according to Ellenberg (1956)

		Class values (or symbols)				
		+	0	1	2	X_h
Class values (or symbols)	+	0	1	0	1	2
	0	0	0	0	3	3
	2	0	2	0	0	2
	3	8	3	6	2	19
	4	1	0	1	0	2
	X_i	9	6	7	6	28

$I(h) = 29.9027,$ $I(h;i) = 56.5515,$

$I(i) = 38.4041,$ $I(h;i) = 11.7554,$

$d(h;i) = 0.7922,$ $R(h;i) = 0.6103.$

Table 3. Contingency table relating species h to species i, based on data in Table 2.

distribution for *Artemisia maritima*, and the row totals represent the same for *Kraschenіnnikovia ceratoides*. The values in the body of the table are the frequencies of the paired observations. The contingency tables for the subsets X_{hA} and X_{hB} are given in Table 4.

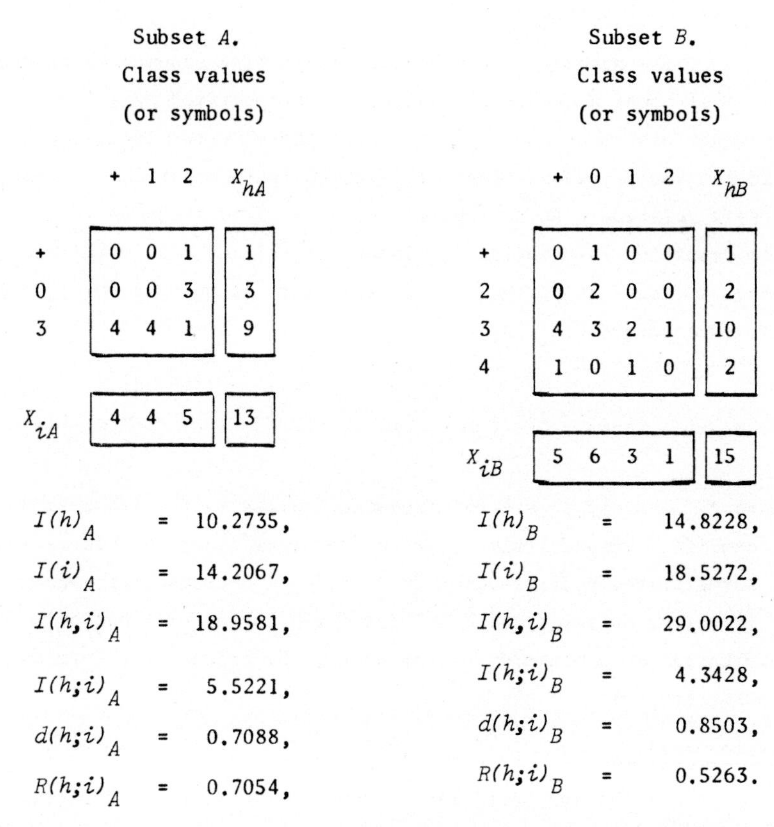

Subset *A*. Class values (or symbols)

	+	1	2	X_{hA}
+	0	0	1	1
0	0	0	3	3
3	4	4	1	9
X_{iA}	4	4	5	13

$I(h)_A$ = 10.2735,

$I(i)_A$ = 14.2067,

$I(h,i)_A$ = 18.9581,

$I(h;i)_A$ = 5.5221,

$d(h;i)_A$ = 0.7088,

$R(h;i)_A$ = 0.7054,

Subset *B*. Class values (or symbols)

	+	0	1	2	X_{hB}
+	0	1	0	0	1
2	0	2	0	0	2
3	4	3	2	1	10
4	1	0	1	0	2
X_{iB}	5	6	3	1	15

$I(h)_B$ = 14.8228,

$I(i)_B$ = 18.5272,

$I(h,i)_B$ = 29.0022,

$I(h;i)_B$ = 4.3428,

$d(h;i)_B$ = 0.8503,

$R(h;i)_B$ = 0.5263.

Table 4. Contingency tables for subsets

Classification

General

The different strategies of classification usually fall into one of **two** basic categories: (1) non-hierarchic sorting of individuals between classes of known parameters, and (2) clustering in a previously unsorted collection with the explicit goal of creating a hierarchy of inclusive classes. The techniques in (1) are related to discriminant analysis, while the techniques in (2) are collectively called cluster analysis.

Non-hierarchic sorting

The problem in non-hierarchic sorting amounts to finding the class among a number of classes, each specified by a frequency distribution $E(X_{i.})$, to which the observed frequency distribution X_h, whose class affiliation is in question, is most closely related. An appropriate measure of relatedness is the discrimination information (Kullback and Leibler 1951; Kullback 1959; Kullback *et al* 1962). This measure is defined for X_h and the ith standard $E(X_{i.})$ by

$$I(h;E(X_{i.})) = \sum_{j=1}^{n_h} f_{hj} \ln \frac{f_{hj}}{E(f_{ij})} \qquad (11)$$

where f_{hj} and $E(f_{ij})$ are the frequency values in the jth cell of X_h and $E(X_{i.})$ respectively. Twice the discrimination information is asymptotically distributed as χ^2 at n_h-1 degrees of freedom. X_h is joined to the class $E(X_{i.})$ for which the discrimination information is a minimum, or for which, the probability corresponding to χ^2 is the highest.

Cluster analysis

Of the many different techniques that could be conceived for information clustering we may distinguish between at least two basic types:

(1) A similarity (dissimilarity) matrix is generated between pairs of rows (or columns) of the data matrix representing the entities to be classified. The information to be used is preferably a relative measure such as Rajski's metric, the coherence coefficient, or the probability corresponding to the mutual information. In the first pass through the similarity

(dissimilarity) matrix entities that are mutually closest are united in pairs. Before the second, or as a matter of fact any subsequent pass, the order of the data matrix is reduced by a number equal to the number of fusions affected, and a new similarity (dissimilarity) matrix is generated between the rows (columns) of the residual matrix. The procedure is continued until the hierarchy is completed. The reduction of the order of the data matrix is achieved by pooling the corresponding values between pairs of rows (columns), representing the entities whose union was indicated by the clustering process.

In some clustering techniques, such as for instance, the average linkage technique of Sokal and Michener (1958), the similarity (dissimilarity) matrix is reduced after each clustering pass by some method of simple, or weighted, averaging (Gower 1967) of the corresponding values between the appropriate pairs of columns, and between the pairs of the equivalent rows indicated by the clustering process.

(2) Some techniques of this group rely on the heterogeneity information defined by

$$\Delta I_{AB} = I(1;2)_{AB} - I(1;2)_{A} - I(1;2)_{B} \qquad (12)$$

where the Is represent the mutual information between two frequency distributions X_1 and X_2 representing the row and column classifications in table A or B in their union $A + B$. Assuming that the data are given as frequencies, the data matrix X can be regarded as a contingency table. Further assuming that the goal of the analysis is a hierarchic classification of the columns of X, the Is in equation 12 represent the mutual information between the column and row classification in A or B, as disjoint subsets of columns, and in their union $A + B$. In agglomerative clustering, A is united with B if ΔI_{AB} is a minimum, or if the probability corresponding to the heterogeneity information is sufficiently large. The probability is defined by the assymptotic relationship

$$2\Delta I_{AB} \simeq X^2 \text{ at } (r_A+r_B-1)(c_A+c_B-1) - (r_A-1)(c_A-1) - (r_B-1)(c_B-1)$$

or $r-1$ when $r_A=r_B$, degrees of freedom. The symbols r_A, c_A, r_B and c_B indicate the number of rows (r) and columns (c) in table A and B respectively. When the clustering is subdivisive, $A + B$ is subdivided if ΔI_{AB} is a maximum or if the corresponding probability is sufficiently small.

The basic data, however, are usually not given as frequencies. It is nevertheless possible to summarize most types of data in frequency tables and then manipulate these tables to derive a measure of heterogeneity between subsets. There are at least two alternative avenues of approach to follow in this connection. The first (equation 13) makes use of the joint information according to equation 3 and the second (equation 14) utilizes an expression related to equation 7 as applied to r dimensional frequency tables. In the case of equation 3, the mutual information is disregarded, while equation 7 represents the mutual information. The decision parameter then is

$$\Delta I_{AB} = I(1,2,\ldots,r)_{AB} - I(1,2,\ldots,r)_A - I(1,2,\ldots,r)_B \quad (13)$$

or

$$I_{AB} = I(1;2;\ldots;r)_{AB} - I(1;2;\ldots;r)_A - I(1;2;\ldots;r)_B \quad (14)$$

where the different Is represent the joint information between the r rows, or the mutual information, within the disjoint subset A or B of the columns of X and in the united subsets $A + B$. It is still assumed at this point that the entities to be classified are represented by the columns of X. A union $A + B$ is accepted if ΔI_{AB} is a minimum. In subdivisive clustering, however, $A + B$ is subdivided if ΔI_{AB} is a maximum.

Concluding Remarks

It should be noted that the different measures of information as presented in this paper, frequently are weighted by the absolute frequencies. Clearly, this dependence on size may not be desired in specific situations when a relative measure of information is appropriate. Measures that qualify in this regard include the average information or entropy (*e.g.* $H(h) = I(h)/n_h$), the probability corresponding to information at given degrees of freedom, or other measures such as Rajski's metric or the coherence coefficient. Furthermore, it must be kept in mind that when the χ^2 distribution is used to approximate the distribution of information the approximation can be improved by increasing sample size. This is a direct consequence of the fact that χ^2 is only a limiting distribution for information.

References

Abramson, N. 1963. Information theory and coding. pp.201. New York, London, McGraw-Hill.

Ducker, S.C., Williams, W.T., and Lance, G.N. 1965. Numerical classification of the pacific forms of *Chlorodesmis* (Chlorophyta). *Aust. J. Bot. 13* : 489-499.

Edwards, A.W.F. and Cavalli-Sforza, L.L. 1965. A method for cluster analysis. *Biometrics, 21* : 363-375.

Ellenberg, H. 1965. Einführung in die Phytologie von H. Walter, Bd. IV, I. Teil: Aufgaben under Methoden der Vegetationskunde. Stuttgart.

Fisher, R.A. 1936. The use of multiple measurements in taxonomic problems. *Ann. Eugenics. 7* : 179-188.

Gower, J.C. 1967. A comparison of some methods of cluster analysis. *Biometrics, 23* : 623-637.

Hartman, H. 1968. Über die Vegetation des Karakorum. Part 1. *Vegetatio, 15* : 297-387.

Jancey, R.C. 1966. The application of numerical methods of data analysis to the genus *Phyllota* Benth. in New South Wales. *Aust. J. Bot. 14* : 131-149.

Khinchin, A.I. 1957. Mathematical foundations of information theory. (Translated by R.A. Silverman and M.D. Friedman). 120 pp. New York, Dover.

Kullback, S. 1959. Information theory and statistics. 395 pp. New York, Wiley, London, Chapman and Hall.

Kullback, S., Kupperman, M. and Ku, H.H. 1962. Tests for contingency tables and Markov chains. *Technometrics, 4* : 573-608.

Kullback, S. and Leibler, R.A. 1951. On information and sufficiency. *Ann. Math. Statist. 22* : 79-86.

Macnaughton-Smith, P. 1965. Some statistical and other numerical techniques for classifying individuals. London, H.M.S.O.

Macnaughton-Smith, P., Williams, W.T., Dale, M.B., and Mockett, L.G. 1964. Dissimilarity analysis: a new technique of hierarchical subdivision. *Nature, 202* : 1034.

Orloci, L. 1967. An agglomerative method for classification of plant communities. *J. Ecol. 55* : 193-205.

Orloci, L. 1968. Information analysis in phytosociology: partition, classification and prediction. *J. Theoret. Biol. 20* : 271-284.

Rao, C.R. 1952. Advanced statistical methods in biometric research. 390 pp. New York, Wiley; London, Chapman and Hall.

Rajski, C. 1961. Entropy and metric spaces. p.41-45 in "Information theory". (ed. by C. Cherry).

Shannon, C.E. 1948. A mathematical theory of communication. *Bell System Tech. J. 27* : 379-423; 623-656.

Sokal, R.R. and Michener, C.D. 1958. A statistical method for evaluating systematic relationships. *Univ. Kansas Sci. Bull. 38* : 1409-1438.

Williams, W.T. and Lambert, J.M. 1959. Multivariate methods in plant ecology. I. Association analysis in plant communities. *J. Ecol.* *47* : 83-101.

Williams, W.T., Lambert, J.M., and Lance, G.N. 1966. Multivariate methods in plant ecology. V. Similarity analysis and information analysis. *J. Ecol.* *54* : 427-445.

Discussion

Q.(Wishart) When you say that the information statistic is approximately distributed as χ^2, do you have to make assumptions about the origin of the sample distribution?

A. It seems that information is a distribution free statistic. If you have a reason to assume that the underlying probability distribution is of a particular kind, for example, binomial or Poisson, then the calculations simplify the analysis. But you don't need to know the underlying probability distribution.

Q.(Tomassone) I often use information theory for statistical problems, but while it can be used with no reference to the distribution, there is a double aspect: firstly, the distribution-free aspect where the additive properties are positive, but secondly, the limit distribution of the information quantity 2I which tends always to χ^2 is only a limit one, and needs some particular restrictions.

A. Yes, that is so.

Q.(Jackson) Are your classificatory sets described at the outset in terms of certain statistical distributions, in which case you take each object and say does this object fit into a class of this type? Do you also take a fixed number of classes and then attempt to add elements to each other?

A. I have some large data, some species and stands, and I classified stands, using the heterogeneity information, into a number of inclusive classes necessary to form a hierarchy. For another application I have done, using the discrimination information, you need to know how many classes you have and the class parameters.

Q.(Jackson) In certain applications, for example information retrieval, such information just isn't available.

A. In most cases it is not available. In ecology, the *á priori* probabilities are very rarely specified, but there may be an instance when they are. You may make a hierarchic classification by some cluster analysis, and find that the classes which are formed at some level are meaningful in respect with some treatments. The problem may arise that if we admit a new sample, what kind of treatment shall we apply to it? The answer is, the one to the class of which the sample in question is most closely related in terms of the discrimination information calculated from the *á posteriori* probabilities. For those of you concerned with probabilistic applications, I think that my field - ecology - is one of the most difficult in this connection. We must rely, in most instances, on deterministic studies and we rarely try to apply statistical tests of any kind.

A MATHEMATICAL STUDY OF THE GENUS PENTREMITES

Charles C. Ostrander

Engineering Experiment Station, Georgia Institute of Technology

In the preparation of this work, I took the view of a geologist; the user of the work of the taxonomist. My training in taxonomy, palaeontology and biology has been superficial, and this is a charitable evaluation.

During my thesis investigation at Emory University, I collected 136 specimens of the blastoid genus *Pentremites* from a small glade near Collegedale, Tennessee. The glade contained a section of some 30 feet of Mississippian (St. Genevieve) strata. An abortive attempt at identification was made using original description (where practical) and other published works. Some specimens were "easily identified" whereas others successfully resisted all attempts at identification.

When I read Galloway and Kaska's (1957) work I thought I had found the key. They had "redescribed and figured" the various species of the genus *Pentremites*, applying numbers to many of the parameters. The list of 133 species existing prior to their work was reduced to 62 American species and 1 doubtful European species. Soon, even this shortened list proved to be overwhelming to me, so I wrote a Fortran program to assist in identification.

This program was written around the figures supplied by Galloway and Kaska (1956). In most instances, they supplied limiting values for the species parameters. In such instances these values were used. In other instances, only a single value was given, *e.g.* a pelvic angle of 140°. In these instances I supplied limiting values based upon their given value and the recommended tolerances stated in their text.

The criteria most stressed by Galloway and Kaska are:

1. Length/width ratio
2. Vault/Pelvis ratio
3. Pelvic angle
4. Convexity or concavity of ambulacra
5. Concavity of interambulacra

In the text describing each species the length and width are given.

The convexity or concavity of the ambulacra and interambulacra are not numerically described. The terms used in their description are convex, concave, concave?, narrow convex, flat, slightly convex, slightly concave, deeply concave, flat, nearly flat, plane, moderately concave, broad v-shaped, and strongly concave. I was at a loss to objectively define, for example, the difference between concave, concave?, nearly flat, slightly concave, and moderately concave so these parameters were omitted from consideration by the computer.

The resultant identifications were appalling. Of 136 specimens whose measurements were submitted to the program, 68 (50%) were reported as unidentified, 47 (35%) were identified as belonging to one of 22 different species (see Table 1). The remaining 21 specimens (15%) had parameters such that they fit into more than one species. (17 of these multi-identified specimens were double identifications, 2 were triple, and 2 were quadruple).

The multiple identifications could (and possibly should) be dismissed as expected since several significant discriminating criteria such as the curvature of the ambulacra were not used. Two glaring facts still remain:

1) 68 specimens were unidentified even though the discriminatory criteria were considerably less rigorous than those of Galloway and Kaska, and

2) 22 different species were identified among the specimens (when including the multi-identifications it is necessary to add 1 more species, for a minimum of 23 different species).

Species	Number of specimens
Pentremites abruptus Ulrich 1917	1
P. basileris Hambach 1880	1
P. biconvexus Ulrich 1917	2
P. chesterensis Hambach 1880	1
P. conoideus Hall 1856	4
P. girtyi Ulrich 1917	1
P. godoni (Defrance) 1819	1
P. godoni abreviatus Hambach 1880	1
P. godoni pinguis Ulrich 1917	2
P. halli Galloway and Kaska 1957	1
P. hambachi Butts 1926	1
P. malotti Galloway and Kaska 1957	3
P. obesus Lyon 1857	4
P. okawensis Weller 1920	1
P. princetonensis Ulrich 1917	1
P. pyriformis Say 1825	3
P. robustus Lyon 1860	1
P. spicatus porrectus Haas 1945	11
P. springeri Ulrich 1917	1
P. tulipaformis Hambach 1903	3
P. tuscumbiae Ulrich 1917	1
P. welleri Ulrich 1917	2

Table 1

Species identified by use of the UNIVAC 1108 Computer

Frequency histograms of 5 parameters (length, width, length/width ratio, vault/pelvis ratio, and pelvic angle) were plotted. Each of the histograms was unimodal. (Plate I). A three phase diagram using length, width, and ambulacrum length has a tendency toward a "bimodal" character (Plate II).

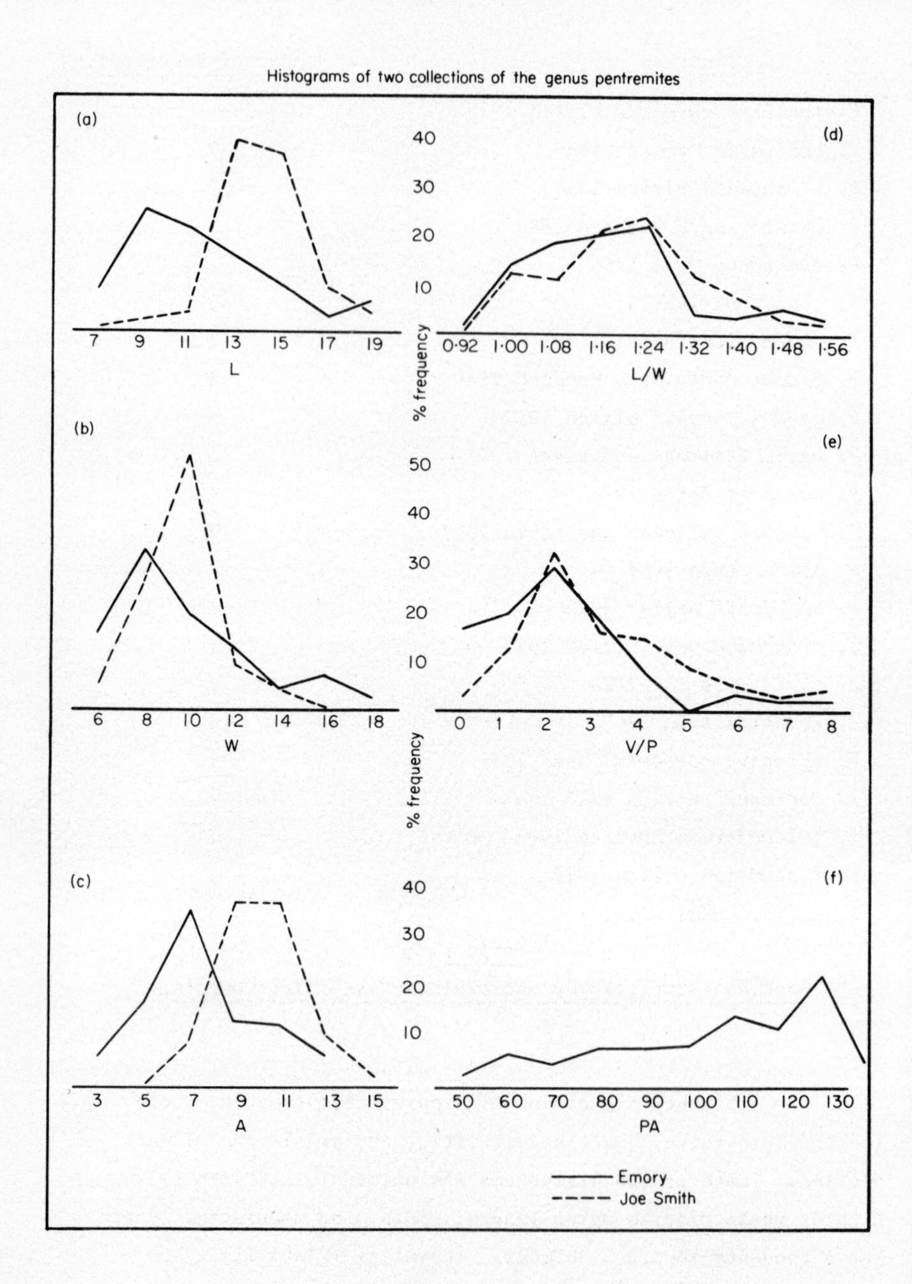
Histograms of two collections of the genus pentremites
(a)
(b)
(c)
(d)
(e)
(f)
40
30
20
10
50
% frequency
7 9 11 13 15 17 19
L
0·92 1·00 1·08 1·16 1·24 1·32 1·40 1·48 1·56
L/W
6 8 10 12 14 16 18
W
0 1 2 3 4 5 6 7 8
V/P
3 5 7 9 11 13 15
A
50 60 70 80 90 100 110 120 130
PA
Emory
Joe Smith

Plate I

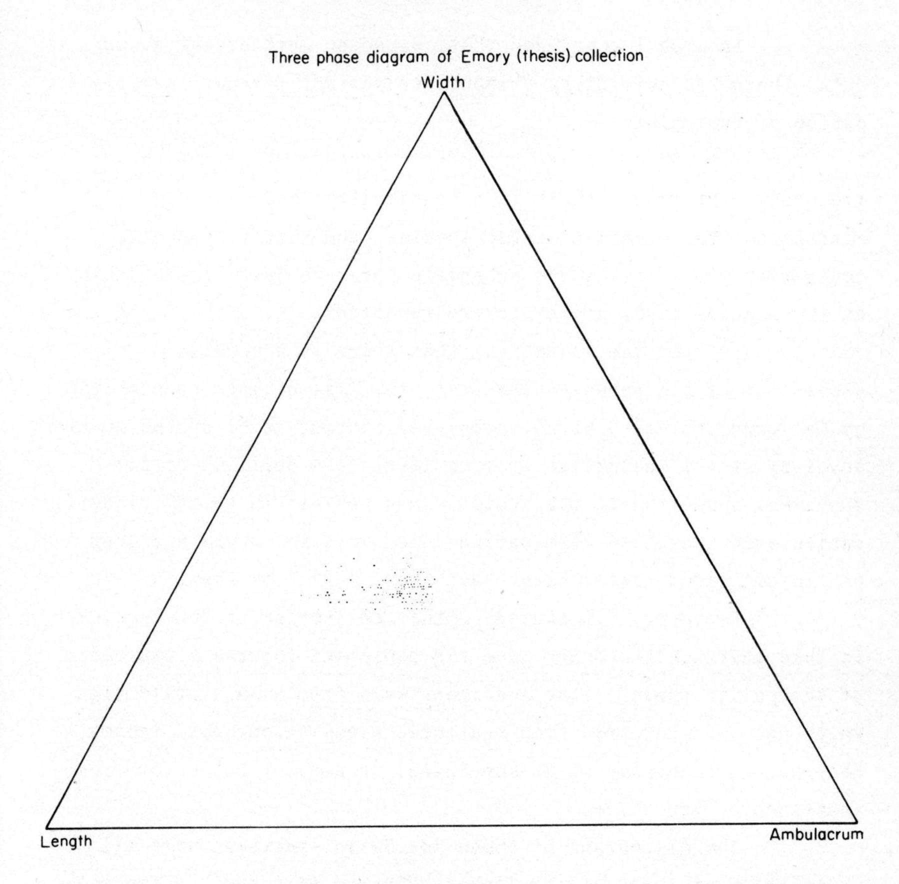

Plate II Three Phase Diagram of Emory (Thesis) Collection

In 1966 I concluded that, based on most of the above data, there was sufficient evidence to warrant further investigation of two points.

(1) Based on the unimodal character of all of the frequency histograms, there is a possibility that the genus *Pentremites* represents a single species, and that the genetic separation should be at the subspecies or race level rather than at the species level as heretofore reported.

(2) On the assumption that there is a genetic separation of the genus *Pentremites*, the list of species compiled by Galloway and Kaska is not complete, whether it be at the species level or at the subspecies or race level. A full 50% of the specimens submitted to the system would not fit in to any classification even though the limitations used were in no way stricter and in all cases more lenient than those listed by them.

Recently, I measured another collection of 206 specimens. In this instance I did not have the equipment for the measurement of the pelvic angle. The specimens were from a bulk collection which had been acquired from scattered areas throughout Alabama, Georgia and Tennessee by Mr Joe Smith, an amateur collector from Chattanooga, Tennessee.

The histograms of these Joe Smith specimens were all essentially unimodal. The 3-phase diagram of these specimens was clearly unimodal in character (Plate III).

A comparison of the graphs of the two collections reveals the following:

(1) The histograms of the length/width ratio and the vault/pelvis ratio (Plate I, d & e) maintain essentially the same modal values for each collection. The character of the curves differ only in detail.

(2) The length histogram of the Joe Smith collection (Plate I a) is more peaked but maintains the same general skewness as that of the thesis collection. The modal value is slightly greater in the Joe Smith collection.

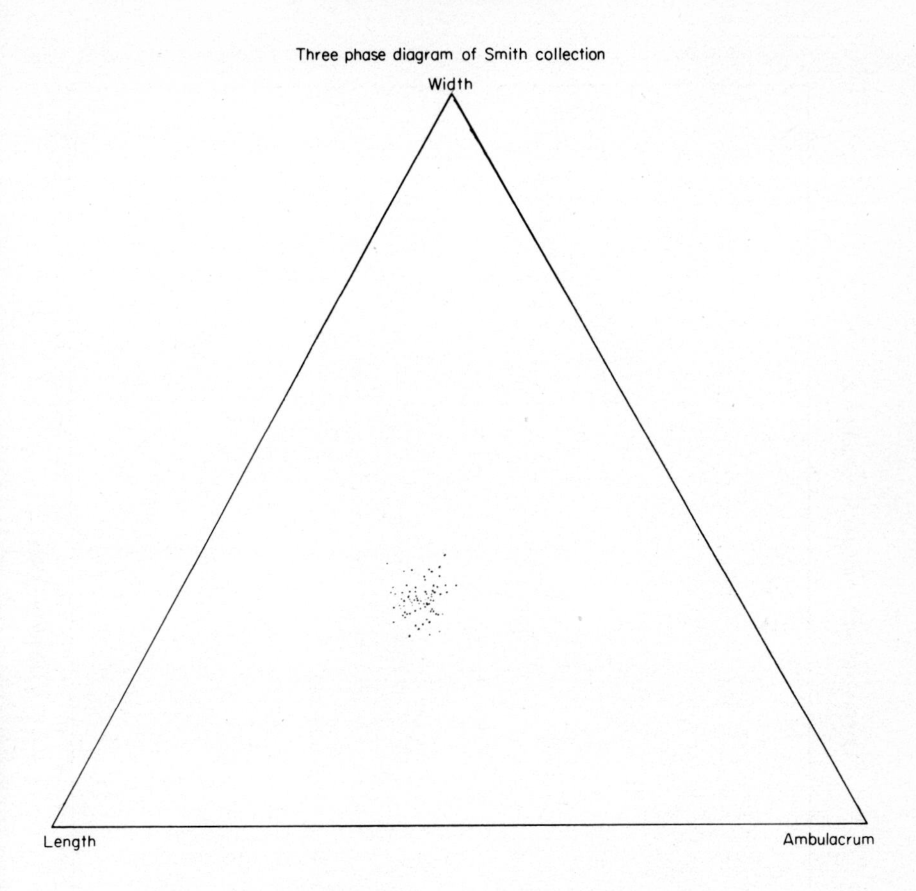

Plate III Three Phase Diagram of Smith Collection

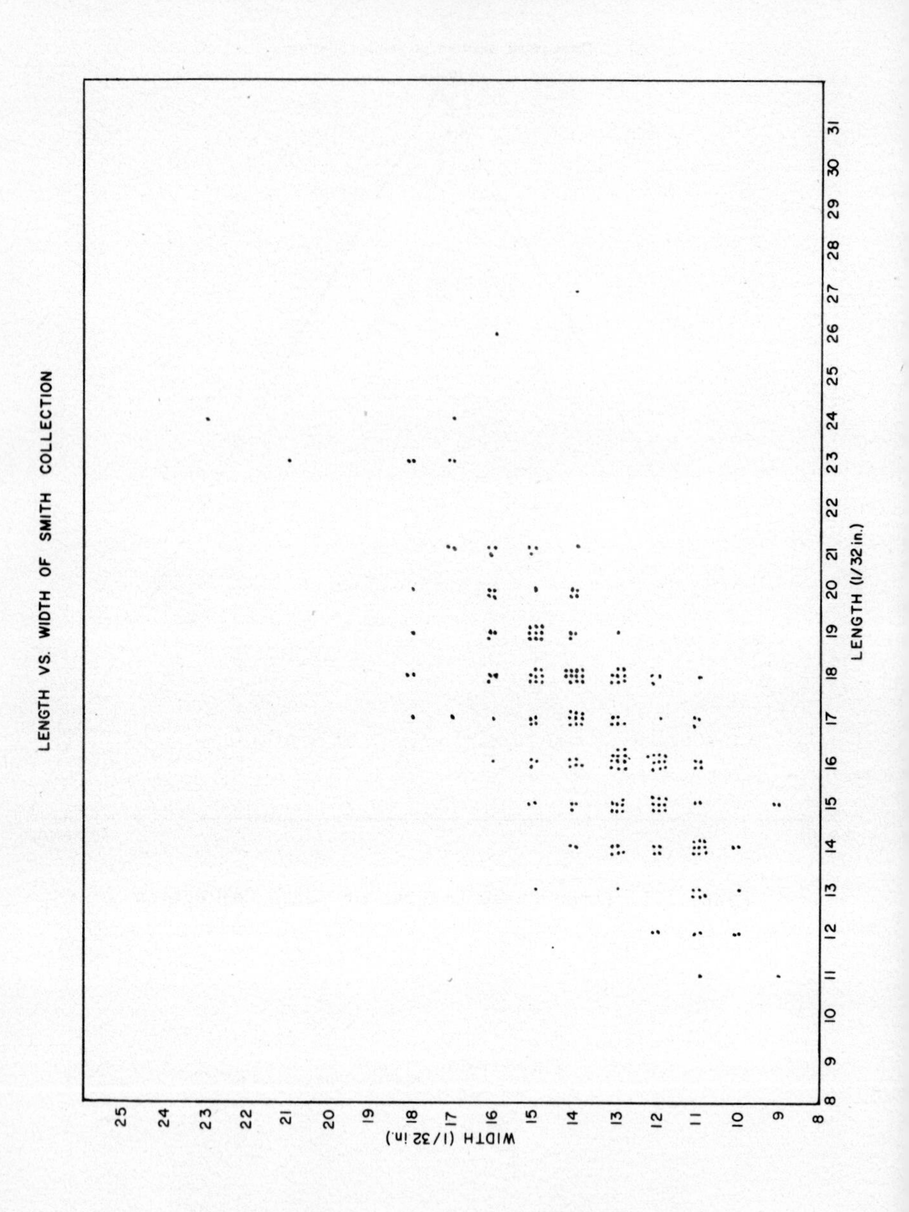

Plate IV Length VS. Width of Smith Collection

(3) The modal value of the width histogram of the Joe Smith collection (Plate I b) is slightly greater. The skewness present in the histogram of the original (thesis) collection is not apparent in the Smith collection. The width histogram is also more peaked in the Smith collection.

(4) The modal value of the histograms of the ambulacral length (Plate I c) is both greater and broader in the histogram of the Smith collection.

(5) A plot of the length versus width of the Joe Smith collection (Plate IV) shows a marked correlation (correlation coefficient = .89). However, the same type plot of the thesis collection shows a quite distinct splitting of the points into two groups (Plate V). (correlation coefficient = .92).

(6) A plot of the length versus the ambulacral length (Plate VI) shows somewhat the same marked correlation in the Joe Smith collection (correlation coefficient = .865). With the thesis collection, the points are once again split into two quite distinct groups (Plate VII). (correlation coefficient = .836).

In looking at the specific specimens making up the smaller (lower) groups of the latter two graphs of the thesis collection I noted that 28 of the specimens fit into both lower groups, whereas only 10 specimens fit into one group but not the other, but even these fit near the fringes. In addition, 24 of these 28 specimens are also present in the smaller mode of the three-phase diagram of the thesis collection and the missing four are on the fringes. The ten fringe specimens above are all within or near the smaller mode of the three-phase diagram.

When the 38 specimens are removed and treated as a separate collection, the larger group has a L-W correlation coefficient of .96, and the smaller group, .964. The L-A coefficients are .962 and .964 respectively.

Marsland (1965, p.543) states:

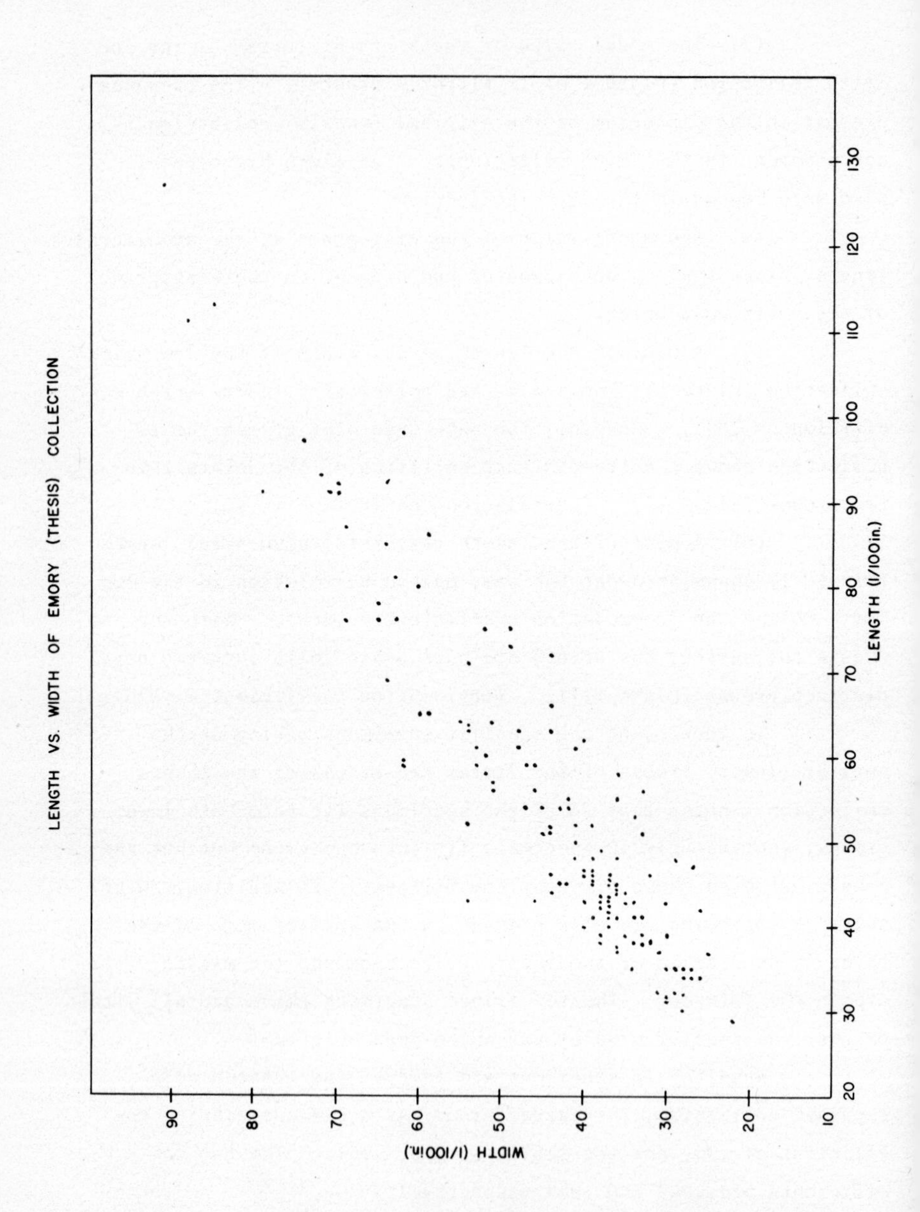

Plate V Length VS. Width of Emory (Thesis) Collection

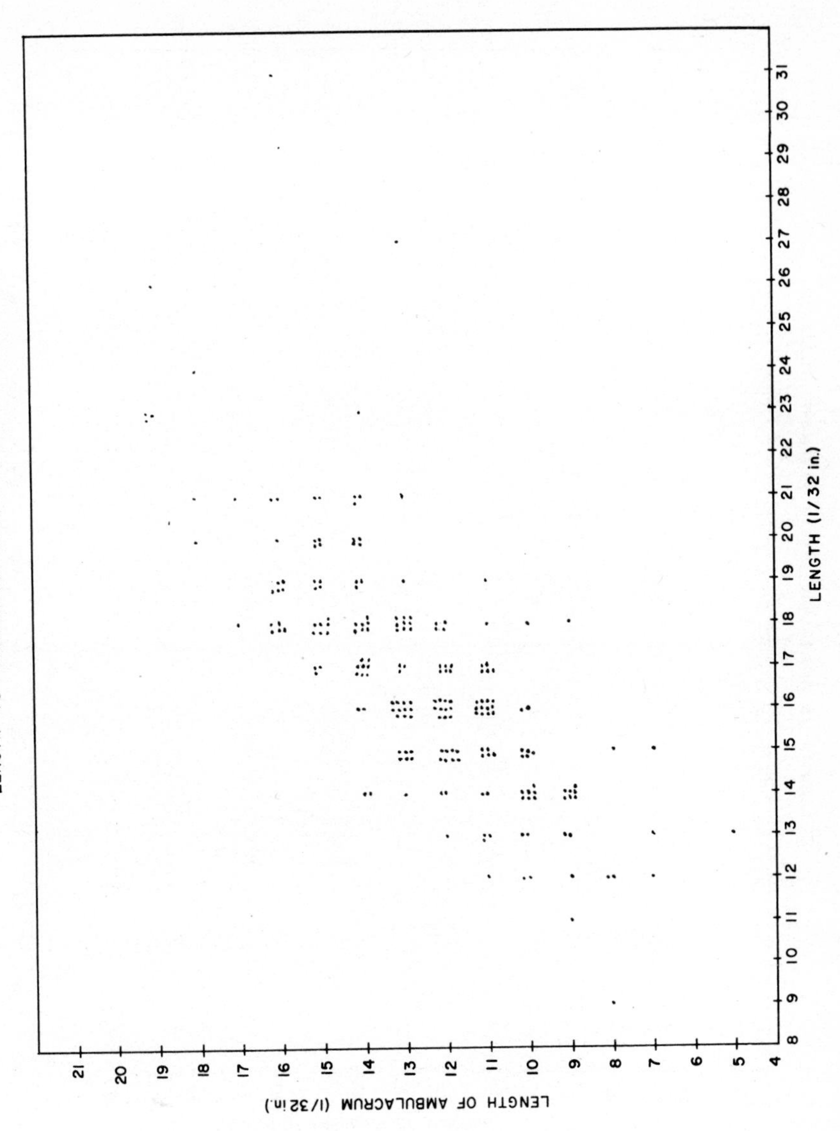

Plate VI Length VS. Ambulacral Length of Smith Collection

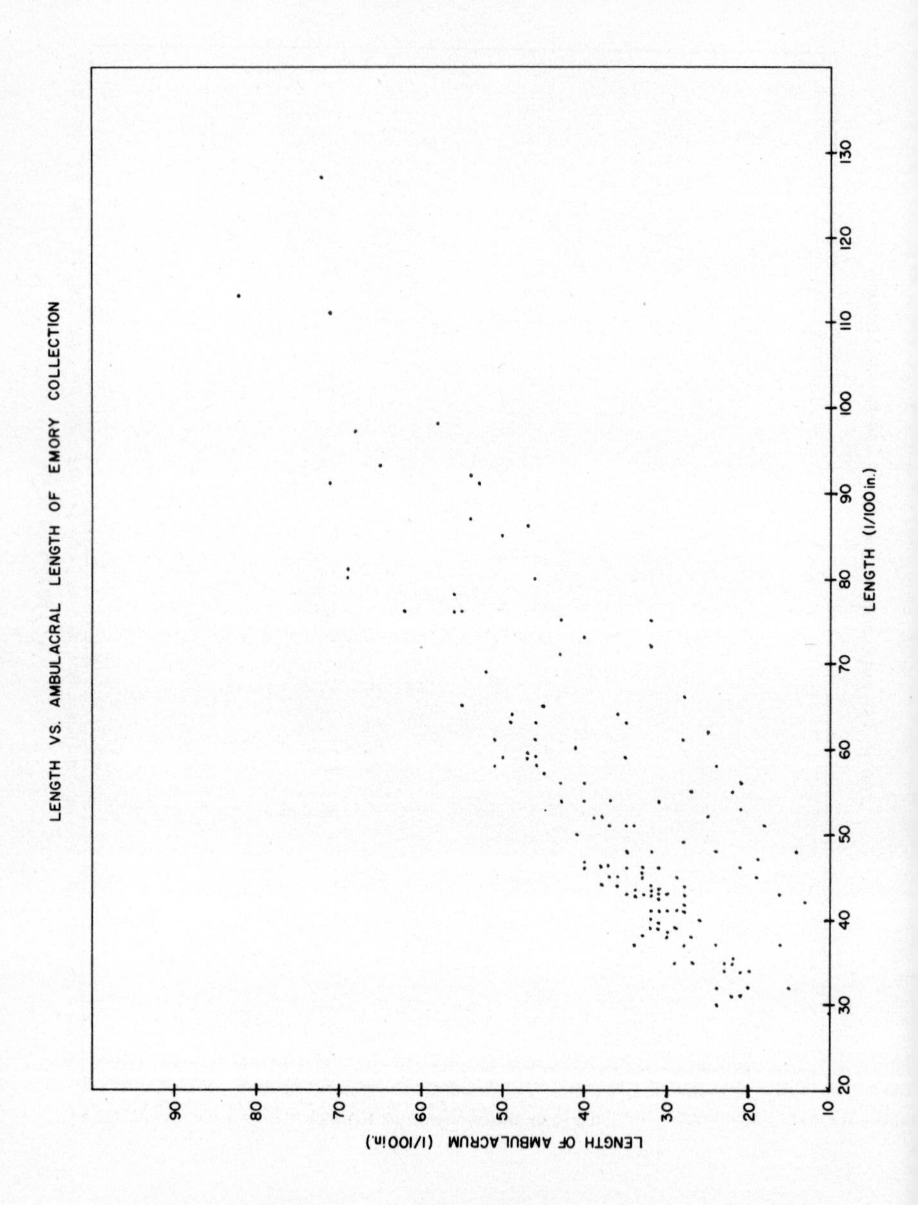

Plate VII Length VS. Ambulacral Length of Emory Collection

> When an environmental change is favorable to a given species, its population will increase, and during this period of less rigorous competition, a larger proportion and a wider variety of offspring will be tolerated for a number of generations . . .When, however, environmental conditions become harder for the species, many of its less fit or weaker varieties may be wiped out, leaving just a few sharply distinct varieties - or incipient new species.

Marsland further states that during the hard times, the species may become extinct over large areas of its former territory, leaving a few sharply distinct varieties, each isolated from the others. With continued isolation and mutation the varieties will become genetically incompatible and may be classed as new species.

Burma (1948, p.741) states that isolation, or "some such factor", is necessary to develop specific differences. Burma further states:

> Usually it is well to make the preliminary assumption that only one species of any genus is present in a sample. In modern marine faunas two very similar species do not commonly exist side by side, although such associations are certainly known.

Based on the above statements of Burma and Marsland, I think that during the Mississippian Period conditions were excellent for the development and survival of many varieties (very possibly this last phrase would be more accurate if changed to: " . . . . for the development and survival of many individual variations . . .") of the genus *Pentremites*. This conclusion is further supported by the abundance of fossilized remains. I am not aware of any evidence to indicate any factor that might have lead to the isolation of any variation or community. The fact of their widespread occurrence throughout Mississippian marine carbonate strata indicates few, if any, isolating factors.

As stated, the 136 specimens involved in the thesis study were collected from one glade in the Ooltewah area about 3/4 mile northeast of Collegedale, Tennessee. They are most likely not from one stratigraphic horizon, although certainly no more than

30 or so feet of section separate the highest from the lowest occurrences.

The possibility of having 22 (more, if the multiple identifications are considered) separate species within such a limited geographical area and stratigraphic interval, when only 63 species have been reported from the entire Mississippian System throughout the world, appears to be remote, particularly in view of the above statement by Burma.

In view of the rather distinct groups appearing in the graphs of the thesis specimens, it appears that genetic separation may have been in progress. It could be at either the species level or at the race level (possibly the incipient species of Marsland). This genetic separation could be a time based factor, but the nature of the collecting area precluded making any such determination.

I further think that during Early Pennsylvanian times the conditions were reversed so that very rigorous competition existed, and finally resulted in the global extinction of the genus *Pentremites* before the advent of the Permian Period.

On the basis of what I have presented, I think that a more vigorous investigation of the genus *Pentremites* is needed to either confirm or deny these preliminary findings. With the present day knowledge of population genetics, this investigation might yield valuable information on the use of genetics and taxonomy in the area of palaeontology. I think that we have been subject to a serious case of overkill in the area of publication of new species with respect to palaeontology.

References

Burma, B.H., 1948, Studies in quantitative palaeontology, I. Some aspects of the theory and practice of quantitative invertebrate palaeontology: *Jour. Palaeontology, v.22*, p.725-761.

Galloway, J.J., and Kaska, H.V., 1957, The genus Pentremites and its species: *Geological Society of America Memoir 69.*

Marsland, D., 1965, Principles of modern biology. Holt, Rinehart & Winston, 4th ed., New York, 723 p.

Discussion

Comment: (Hall) — Where there is an apparent continuum, it may be appropriate to set up "Type" points throughout the range, for provisional reference. These may be helpful in later stages of a project.
Was there any appearance of clustering in the material (disregarding the numerical evidence)?

A: The extremes look like separate species, but there is a complete gradation between the extremes. The size differences are probably age differences.

Q:(Parks) You apparently used only three characters, length, width and length of ambulacrum. Were there not other characters that could be used?

A: Yes, but I had already lost half of the identifications, and if more characters were used I would have lost more still.

Comment: (Parker-Rhodes) There is great value in the use of Numerical Taxonomy for showing-up cases of bogus species-making, which occurs in living biology as well as in palaentology.

Q:(Walton) Is Dr Ostrander satisfied that in rejecting the many non-numerical characters which were used in previous classifications he is not losing some essential characteristics of the organism? Do not his results simply reflect the restricted number of parameters he has used?

A: Yes, there are many other characteristics that should be used, even internal organs are preserved in some rare specimens. However, because of their rarity it is difficult to use them as necessary discriminant characters.

Q:(Wishart) On the question of concavity, why not use volume as a crude measure?

A: Fossils are frequently fragmentary, so that volume would not be a practical measurement.

AUTOMATIC CLASSIFICATION IN THE ECOLOGY OF THE HIGHER FUNGI

A. F. Parker-Rhodes & D. M. Jackson

Cambridge Language Research Unit

While much work has been done on the use of automatic classification in construction of general-purpose classifications of various groups of organisms, such techniques may well show to better advantage in producing special purpose classifications. It often happens that the principles used in such work are ill suited to handling by the intuitive methods which still play the major part in conventional taxonomy; it is in such cases that automatic classification methods may be looked to for a relatively effortless result.

The work reported in this paper relates to the data on the ecology of higher fungi (Basidiomycetes) collected over a period of 16 years in a variety of localities, mainly woodland, in the neighbourhood of Flatford, Suffolk. This data is particularly good material for automatic classification for the following reasons: (1) in its original form, consisting of records on hand-written index cards, it would be a very formidable task to extract any kind of ecological classification from it by hand; (2) it can readily be presented under two categories of descriptors, namely the occurrence of a given species in a given locality, and its occurrence in a given year, leading to different but related classifications, which one might characterize as 'ecological' and 'meteorological' respectively; and (3) the quality of the data is low, in the sense that it results from pure observation, unguided by any experimental design other than that of repeating the observations at the same time of year, and is thus not well suited for the usual techniques of statistical analysis.

In principle, automatic classification should be able to produce results which, by conventional procedures, would require an amount of computation-time disproportionate to the quality of the expected output.

Previous Work in the Field

To the best of our knowledge, no application of automatic classification techniques to data comparable with ours has been published.

Methods Used

The computational methods were those developed by the Cambridge Language Research Unit, originating in the work of Needham and Parker-Rhodes [1,2]. In its present form the method is embodied in a program package operating on the Cambridge University Titan computer. The work done may be divided in to two stages. The first stage is the calculation of the similarity matrix from the original data to give an estimate of the similarity between each pair of objects on the basis of the properties they exhibit. The second stage is the extraction of classes of objects* on the basis of individual similarities between objects as provided by the matrix, and makes no direct use of the original data. The classification procedure itself is an iterative approximation or 'seed' (which we take to be a single object), up to a set of objects stable by the particular criterion adopted. Each 'seed' thus leads to a partition of the universe of objects of which the smaller moiety is considered to be the class sought. The essential operation is to perturb the iterand by transferring a single object across the boundary, and leaving it on that side which satisfies our criterion best. The criterion depends on three quantities associated with any set of objects, namely (a) the sum of the similarities of all pairs of objects within the set,

* We use the term 'class' throughout for the aggregates of objects we deal with. There is urgent need of agreement on terminology here: hitherto we have used the term 'clumps' for the particular kind of classes our methods produce, but a more general term is to be preferred in view of the wide variety of techniques now in operation. Although the word 'class' has a special use in biology, and some terminological strings attached to it in logic, it seems, for the present, to be the best choice.

(b) the sum of similarities between the members of the set and its complement, and (c) the number of objects in the set (*i.e.* sum of self-similarities). We define a function, called the 'cohesion' function, of these three arguments, which measures the overall likeness between the set and its complement. Our criterion is then that the cohesion should be minimised. Each putative partition is then perturbed in respect of every object, and if none of these perturbations reduce the cohesion, the partition is accepted as defining a useful class. Different classifications are obtained on a given set of data by varying the similarity and cohesion functions; those used in the present work are described in the section on 'Calculations' below.

The Data

The data on which this work was based consisted of the accumulated records of a series of 'fungus weeks' held at the Flatford Mill Field Study Centre of the Field Studies Council, from the years 1952 to 1967 inclusive. The earlier years of this series have formed the subject of an earlier paper by one of us [3], in which conventional statistics were used. The records are kept in the form of 8 x 5 index cards, one assigned to each species or variety, on which every observation of the given species within the Flatford area is recorded. Identifications are due to A.F. Parker-Rhodes, except in 1956 when Dr F.B. Hora was in charge. All the observations were made in the last (Wednesday-bounded) week beginning in September. The number of persons participating has varied considerably, from 3 to 15, averaging about 8. Evidence has been given [3] that this variation is not reflected to any great extent in the number of species recorded.

The manner in which such observations are made is not conducive to statistical sophistication. While a positive record can be accepted as sound evidence that a given species did occur on a certain year at a certain place, absence of a record is not evidence of non-occurrence. While most of those taking part are students, and not qualified to identify most of the species encountered, there have been in most years more than one competent mycologist present, so that errors of identification are likely to be negligible. But the only source of information in the records as to the ecological circumstances of each observation is that subsumed by the locality itself. Since most localities are anything but pure stands, or even typical specimens of well-known plant associations, the amount of noise in the data is considerable. The first question to be answered is then whether our methods of treatment of the data are powerful enough to extract a 'signal' from the background of noise. That is, can we find reasonably well-defined ecological classes by taking the localities in which they have been observed as the only descriptors.

Of the original records, referring to over 800 species and varieties and some 25 localities, a selection of 18 localities was made, by excluding all those which were only visited once and produced less than 20 records. Next, all those species were rejected for which there were less than 5 records in accepted localities; this reduced the number to less than half. Finally, varieties (and some species of controversial taxonomic status) were merged with the nearest good species. This gave us a list of 349. An exception was made in the case of *Laccaria laccata*, whose four commonest colour variants were separately listed, though they probably involve alleles at only one or two loci: the purpose of this was to see whether they would be separated in any consistent manner, a result which would tend to counter the suggestion that they are balanced polymorphs. In the result, no answer was obtained. No other comparable species produced enough records for all its variants, though *Amanita citrina* could perhaps have been treated in this way.

The list of species, with a coded summary of their records, is given in Appendix I. The Localities are described briefly in Appendix II.

The Calculations

From the raw data two sets of descriptors were extracted, referring to localities (L) and years (Y). These are the descriptors given in Appendix I. By 'objects' we shall hereinafter mean species of fungi (with the caveats given above). We thus start from two object-property arrays, OP/*L* and OP/*Y*.

In choosing the similarity function to use, out of the many now available (see, for instance, Ball [4]), we must be guided by the nature of the data. In our case, the situation is that a record carries information, whereas the absence of a record does not. That is, the fact that *Clitocybe odora* was found in Dodnash Wood in 1957 means that species did indeed occur there at that time; but the fact we have no record of this species in any other year in that locality does not mean that it did not occur in any other years. (It should perhaps be mentioned that in general Basidiomycetes perennate as mycelia in the soil or other substrate; 'occurrence' in this context means the production of fructifications. The presence of non-fruiting mycelium may be suspected but can very rarely be demonstrated). The most appropriate similarity function in this situation is that introduced by Tanimoto [5]:

$$S(i,j) = \frac{N(k \mid d(i,k) \wedge d(j,k) \neq 0)}{N(k \mid d(i,k) \vee d(j,k) \neq 0)}$$

where $d(i,k)$ is the observation (presence or absence) recorded for object i, property k, in the OP-array d. N denotes the cardinal of the set of entities specified by the criterion given.

The computation of the similarity matrix, S, whose components are the $S(i,j)$ defined above, presented difficulties on account of its size. The method adopted allows for the possibility that the matrix may be larger than the amount of store available in the computer, and allows it to be calculated using any amount of store the user may specify; this method is described by D.M. Jackson [6]. In our case, the time taken was 20 minutes, using 8K of 48-bit words of core store. The density of the matrix (that is, the proportion of components different from zero) was about 0.9; if the matrix, without its zero components, were to be stored all at once it would have occupied 56K words. This bulk can be reduced by thresholding, that is by selecting an arbitrary value of $S(i,j)$ and equating all actual values below this to zero and leaving those above it unchanged. The choice of threshold is evidently affected by considerations arising from the computer facilities available; in our case, the threshold used was 0.29, and the matrix after thresholding occupied 17K words.

Two cohesion functions were used on both OP-arrays. These were defined as follows:

$$G_1(a) = S_{ab}^2 / S_{aa} S_{bb} \quad *$$

$$G_2(a) = \left[\frac{S_{ab}}{S_{aa}} \right] \left[\frac{N_a(N_a-1)}{S_{aa}} - \frac{S_{aa}}{P.N_a(N_a-1)} \right] \quad **$$

Here the partition referred is into two complementary sets a, b of which a is the putative clump. S_{xy} is the sum of the similarities between members of x and of y. N_a is the number of objects in a, and P is an arbitrary parameter. $G_1(a)$ is designed to find classes whose complements are as well defined as themselves, that is, to find good partitions of the universe. $G_2(a)$ allows the internal similarities of a and the separation of a from b to be adjusted relative to each other by the parameter P. With both cohesion functions the average time to extract from the similarity matrix one class was about 0.75 sec.; the number of classes extracted was 349, but of these many were identical with others, and the actual number of distinct classes varied according to the conditions imposed on the calculation.

* function due to A.F. Parker-Rhodes

** function due to D.M. Jackson

To assist in the interpretation of the output from the computer, a number of tables were generated, besides the basic list of classes with their contained objects. These included: (a) a list of classes omitting repetitions; (b) a list of objects with the classes each occurred in; (c) a list of classes with the numbers of its objects exhibiting each of the properties (property-profile); (d) lists of objects having the same sets of classes containing them; (e) the same as (a) but with numeral references replaced by the full name of the fungus species. The time taken to calculate this information was about 1 min. 25 secs.

The Results and Their Interpretation

We give first the results for the classification by localities, since these results can be compared with what is in general known about the ecological preferences of the species concerned and an estimate formed of the reliability of the data under this kind of analysis. Afterwards we discuss the classification by years, the significance of which has no comparable external check.

The symmetric cohesion function G_1 had the following effects:

(a) the classes produced were numerous and exhibited strong overlaps;

(b) well under half the species were included in any of them; and

(c) when two classes were placed in the same superclass whenever their overlap reached 10 p.c. of their joint membership, the number of superclasses was fairly small, and corresponded well with accepted ecological classification of the fungi.

The asymmetric cohesion function G_2 had these effects:

(a) the classes produced had little overlap, and most of them were well-characterized ecologically;

(b) about half of the species fell into classes of 12 members and above, and about three quarters of them into classes of 6 members and above;

(c) there was no obvious internal criterion by which to unite classes of similar ecological complexion, but if those classes were grouped together which had overlaps with one and the same G_1 - class, a good classification was obtained.

Three of the 18 localities (Dodnash, Staverton, Commons Wood) have a particularly diverse floral composition, and correspondingly produce the greatest numbers of species (see Appx.II). If the records from these localities are expunged, the results are markedly improved for both cohesion functions.

If we consider only classes with 12 members and over, there is a marked negative correlation between the species classified by the two cohesion functions. This effect disappears if classes of down to 6 members are included. The great majority of species classified in large classes by both cohesion functions are ones with well-marked ecological preferences (see Appx.III).

The principal ecological classes which can be extracted from the automatic classification are as follows. (a) Grassland fungi, with a less well-defined subgroup of heathland species; (b) fungi associated with birch (this appeared distinctly only in the G_2 classification); (c) fungi of coniferous woods; (d) those of deciduous woods, including two less distinctive components, associated with beech and oak respectively. In the G_1 classification species occurring in mixed coniferous-deciduous localities tended to be classified with the former, making (c) the largest class; this bias did not occur with the G_2 classification. No specifically 'mixed-wood' classes appeared to be produced by either.

The reliability of the classification, in relation to the ecological characterization of the species usually found in text books (which are of approximately the above degree of fineness), was estimated by considering what proportion of the species were found in classes whose ecological character, as judged by the majority of members, was unlike that of the prepositus. In the G_2 classification, 75 p.c. of the species are classified; of these, 85 p.c. occur in classes with the expected ecological character, while 5 p.c. occur in a narrower range of classes than one might expect. Of the 10 p.c. which occurred in classes of the wrong character, 2 p.c. turn out to be the result of single anomalous records. This leaves a residue of 8 p.c. of the species which are classified at all which are plainly classified wrongly by our procedure. Having regard to the still inadequate knowledge of the ecology of the higher fungi generally, this is perhaps as good a result as could be hoped for. A certain amount of noise must be expected to remain when data such as ours are subjected to any sort of analysis, though one could have wished to reach the magic threshold of 5 p.c. There was no correlation between mis-classification of a species and the total number of records for it.

Some correlations were found between the classes extracted by our procedure and taxa of the accepted general-purpose classification of the fungi; these occurred only with the G_2 classification. The genus *Clitocybe*, with 15 species, had 5 in one class of 12, and 3 each in two classes of 13 and 15, giving a probability of the null hypothesis less than 0.001; other cases with better than 5 p.c. significance were *Collybia* with 7 out of 13 species in one class of 17 and the *Clavariaceae* with 3 and 5 out of 9 species in two classes of 13. A borderline case was that of the *Russulaceae*, with 5 and 7 species out of 42 in two classes of 12 members each. These cases suggest that there is a tendency, albeit not very strong, for taxonomically related species to resemble each other more in their ecological character than unrelated species do. This result is, perhaps, hardly surprising; but it is of some interest in view of the fact that ecological characters are used in conventional taxonomy hardly at all above the species level.

The Classification by Years

When the same procedures were applied to the data contained in the array OP/*Y*, the following results were observed: (a) the output from the cohesion function G_1 was similar in type to that produced from the array OP/*L*, but an even lower proportion of the species were classified (86 altogether); (b) using the function G_2, no output was obtained, on account of incompatability of the input with the assumptions underlying the algorithm.

On closer inspection of the output from the function G_1, it was found that, though the classes could be sorted into superclasses using the same overlap criterion which was used above, the resulting superclasses were (a) of very unequal size (b) the largest ones were characterized by descriptor profiles disconcertingly similar, (c) they failed to segregate the members of the *Polyporales* from the *Agaricales*, notwithstanding the known difference between these orders in their reaction to meteorological factors. The one positive indication was that one of the superclasses contained 10 species of the *Russulaceae* as against 10 of the other families; this is a highly significant correlation. The superclass in question was characterized by exceptionally regular fruiting, which is admittedly a property of many *Russulaceae*; this result is a genuine correlation of the type looked for, but a very special case.

To obtain further light on this situation, the following data, obtained from Monthly Weather Reports of the Meteorological Office [7], were tabulated for each of the years of the study; Rainfall in September, Rainfall in July-August-September, coldest monthly temperature in preceding 12 months, hottest monthly temperature in preceding 12 months. One would expect that, if the classification by years was in any sense a classification by meteorological factors, years giving similar profiles of records over the above-mentioned superclasses should also have similar descriptions meteorologically. No such correlation was found.

It would seem legitimate to conclude that, whereas the use of our localities as ecological descriptors, crude and confused though they are by normal classification standards, yields a fairly satisfactory classification, the analogous use of years as meteorological descriptors involved an input of noise greater than our classification techniques can cope with. Though this is a negative result, it is not without interest that we should, accidentally, have hit upon an important threshold effect in the application of automatic classification to questionably appropriate data. The effect of these techniques is to extract a signal from a background of noise; like any procedure with this effect, there is a limiting level of noise which leads to loss of the signal. There have in the past been some failures, and (naturally) many more successes, reported in the literature on automatic classification; in this case we have one of each, side by side. Obviously, this is not the only possible cause for failures of the technique, but it must be one of the most important ones.

General Commentary on the Results

What, taken together, do our results show as to the utility of automatic classification as applied to the kind of question discussed here? Clearly, the kind of classifications which emerge from our procedures are not wholly satisfactory as a prime source of information on the ecology of the fungus species concerned. Many individual species are mis-classified, and some of the classes are hard to characterize at all. While no quantitative data are here presented in support, it is probable that the remarks on the ecology of individual species in standard textbooks are in error for about the same proportion of cases (say 15 p.c.) as ours are; but this is a comment on the backward state of fungus ecology, not on the excellence of the present contribution. The basic reason for the defects of our output is simply that the input does not consist of ecological descriptors of the best available kind, but of data only indirectly correlated with this. It is remarkable, perhaps, that anything recognisable should come from such material at all. What we can claim to have demonstrated is the power of automatic classification to extract sense out of low quality data. Had the input consisted of ecological observations of the usual detailed kind, there can be little doubt that the same procedures could have delivered a classification actually superior to any presently available for this group of organisms. It is likely that someone is already in possession of such data, which could be so used.

References

1. Parker-Rhodes, A.F. Contributions to the Theory of Clumps. *C.L.R.U. Workpaper M.L. 138* (1961).

2. Needham, R.M. Theory of Clumps II. *C.L.R.U. Workpaper M.L. 139* (1961).

3. Parker-Rhodes, A.F. The Statistics of Fungus Forays. *Trans. Brit. Myc. Soc. 38*, p.283 (1955).

4. Ball, G.H. Data Analysis on the Social Sciences: What about the details. *Proceedings of the Fall Joint Computer Conference 1965*.

5. Tanimoto, T.T. An Elementary Theory of Classification and Prediction. I.B.M., York Town Heights, N.Y. 1958.

6. Jackson, D.M. A note on a procedure for generating large similarity matrices for use in automatic classification. *C.L.R.U. Workpaper M.L. 194*, (1967) (mimeo).

7. Meteorological Office. Monthly Weather Reports. *H.M.S.O.* 1952-1967.

8. Dennis, R.W.G., Orton, P.D. and Hora, F.B. A new Check List of British Agarica and Boleti. *Trans. Brit. Myc. Soc. 43.* Suppl. (1960).

Acknowledgements

We would like to express our indebtedness to the Director and Staff of the Cambridge University Mathematical Laboratory for making available their facilities and for sundry help in Processing the material. We would also wish to thank Mr F.J. Bingley the Warden of Flatford Mill Study Centre for the loan of the records, and of course all those who over the years have participated in collecting the data.

Discussion

Q. (Hall) Does the clump theory system have a means of computing quantitative abundance data? Abundance is an important consideration in ecology and with presence/absence data alone much information is lost.

A. I would freely agree that we are losing lots of precious information - it is not so much "losing" but in fact never having it. Of course if you have got it so much the better. The point is how much can you do if you have not got it? Habitually a good, down to the wood ecologist would look at it and say it was a load of rubbish but in fact there is a lot to be got out of it. With regard to "if we could cope with it if we had it", I think you suspect that we did have it but just did not use it. (loud laughter).

Comment: (Hall) I am sure you did the best job you could.

A. If you do have quantitative data yes indeed one can cope with it - it is quite a lot more work at the computing end because you have to think of something else besides yes and no, but it can be done.

Q. (Davie) Do the clumps depend on the order in which objects are selected for testing for inclusion. If so have you tried different orders?

A. I think that answer is no but I would like Mr Jackson to answer that one.

Comment: (Jackson) I think that the unfortunate fact is that the process is order dependent. It is not something I like very much. Randomised permutations have been tried and significantly different results were not obtained. Further than that I would not like to say.

Q.(Crawford) If there is no correlation between clusters and meteorological variation from year to year can one be found between habitat and annual weather variation?

A. The answer to that is that we have not tested for this. This would be possible to do and would be an interesting question but we have not looked into it.

Q.(Davie) What theoretical considerations led you to the use or the invention of these functions G_1 and G_2. I think G_1 is a fairly obvious one to try but G_2 is not so obvious - certainly not to me.

A. Well Dr Jackson invented it, perhaps he will tell you.

Comment (Jackson) Concerning the formulation of the cohesion function G_2. The cohesion function consists of two factors. The left hand factor measures, informally speaking, the strength of the embedding of the putative class in the population. The other part, which may be adjusted by means of the parameter P concerns the tightness of the internal structure of the class. The purpose of P is to adjust the effect of one factor on the other while the class is being formed. It should be noted that

$$\frac{N_a(N_a-1)}{2}$$

is the least upper bound of S_{aa}. The cohesion is not very sensitive to the value of P and in reality we use only two values namely 1 and 2. In the limit as P tends to infinity we arrive at another cohesion function, one on which some tests have been made.

Comment: (Parker-Rhodes) I might perhaps mention that the time taken to calculate these is really very short - it looks complicated but the total calculation comes out at about $1\frac{1}{2}$ seconds/class.

Comment: (Jackson) 0.75 seconds in fact.

Comment: (Parker-Rhodes) That's even more reasonable.

APPENDIX I

List of Species to be Classified

After the name of the species, the next two columns give the vector descriptions for localities and years respectively. Each vector is in octal digits. The reference of each bit of each digit for *localities* is given in Appendix II. For years, the 18 bits in the 6 octal digits refer to the years 1950 to 1967 in order (the 1st two being always zero). The fourth column indicates the total number of records for each species, divided by 5 neglecting fractions.

TREMELLALES			
Auricularia auriculajudae (Porta ex.Fr.) Mer.	462226	173777	7
A. tremelloides Gray	462062	172761	4
Calocera cornea (Batsch ex. Fr.) Fr. + C. striata, C. fasciatoramosa	740062	163324	2
C. viscosa (Pers. ex.Fr.)Fr.	004232	020064	1
Dacrymyces deliquescens (Bull. ex.Fr.) Duby	766272	173725	6
POLYPORALES			
Athelia centrifuga (Lev.) Eriks.	757732	110757	8
Auriscalpium vulgare Gray	004212	001377	2
Bjerkanders adusta (Willd.ex.Fr.) Karst.	072212	022451	2
B. fumosa (Pers.ex.Fr.) Karst.	240013	002660	1
Coltricia perennis (L.ex.Fr.)Gray	004000	003051	1
Coriolus hirsutus (Wulf.ex.Fr.) Quel.	660032	011532	2
C. pubescens (Shum.) Quel.	661142	171640	1
C. versicolor (L.ex.Fr.) Quel.	762372	137713	7
C. zonatus (Fr.) Quel.	400302	140214	1
Daedalea quercina (L.ex.Fr.) Fr.	762572	137777	8
Daedaleopsis confragosa (Bolt. ex.Fr.) Schrot.	740072	173773	5
Elfvingia applanata (Pers.ex.Fr.) Karst.	022212	027776	5
Fistulina hepatica (Huds.ex.Fr.)Fr.	472332	173677	8
Fomitopsis annosus (Fr.) Karst.	014216	075777	6
Gloeostereum purpureum (Pers.ex. Fr.)Donk	566012	053361	2
Grifola frondosa (Dicks.ex.Fr.) Gray	040050	010130	1

G. gigantea (Pers.ex.Fr.) Murr.	061212	033777	3
Hapalopilus nidulans (Fr.) Karst.	042222	030003	1
Heteroporus biennis (Bull.ex.Fr.)Laz.	220022	104340	1
Hirschioporus abietinus (Dicks.ex. Fr.) Donk	004232	010436	1
Hymenochaete rubiginosa (Dicks.ex. Fr.) Lev.	442132	173776	6
Inonotus hispidus (Bull.ex.Fr.) Karst.	062633	171572	3
Laetiporus sulphureus (Bull.ex.Fr.) Murr.	360062	030533	3
Merulius corium (Pers.) Fr.	260422	102640	1
M. tremellosus (Schum.) Fr.	460010	162323	2
Phellinus torulosus (Pers.ex.Fr.) Boud.	060000	000216	1
Phelbia radiata Fr.	140002	000306	1
Piptoporus betulinus (Bull.ex.Fr.) Karst.	554272	172337	5
Polyporus squamosus (Huds.) Fr.	162261	172572	5
Polyporellus picipes (Fr.) Karst.	763010	022767	2
P. varius (Fr.) Karst.	040600	040031	1
Schizophyllum commune Fr.	520000	150410	1
Stereum gausapatum Fr.	762162	153776	6
S. hirsutum (Willd.ex.Fr.) Gray	577773	177777	25
S. luteocitrinum Sacc.	640060	000360	1
S. rugosum Pers.	742162	120777	3
S. sanguinolentum (A+S.ex.Fr.) Fr.	762176	173367	7
Thelephora intybacea (Pers.) Fr.	000210	000324	1
T. terrestris Ehrh. ex. Fr.	044202	163367	3
Trametes gibbosa (Pers.) Fr.	060010	010740	1
Tyromyces caesius (Schad.ex.Fr.) Donk	342162	123363	2
T. chioneus (Fr.) Donk	010312	012022	1
T. fissilis (Bk. + Curt.) Donk	440022	103320	1
T. fragilis (Fr.) Donk	444110	041600	1
T. lacteus (Fr.) Donk	760332	076265	3
T. semipileatus (Peck) Murr.	640020	001712	1
Vuilleminia comedens (Nees.ex.Fr.) Mre.	052062	001776	3
Xylodon versiporum (Pers.ex.Fr.) Gray	777777	177777	31

AGARICALES

Russulaceae

Lactarius (Lactarelis) piperatus (Scop.ex.Fr.) Gray	100120	113000	1
L. (L.) vellereus (Fr.) Fr.	100132	001277	2
Russula (Lactarelis) adusta (Pers.ex.Fr.) Fr.	650522	160527	4
R. (L.) delica Fr.	520530	171070	2

R. (L.) nigricans (Bull.ex.Mer.)Fr.	750132	137377	7
Lactarius blennius (Fr.ex.Fr.) Fr.	204110	132367	2
L. camphoratus (Bull.ex.Fr.) Fr.	140132	025067	2
L. decipiens Quel.	200102	001061	1
L. deliciosus (L.ex.Fr.) Gray	004202	010227	2
L. glyciosmus (Fr.ex.Fr.) Fr.	564372	177377	12
L. helvus (Fr.) Fr.	044050	073220	1
L. hepaticus Plowr.	154052	172157	3
L. mitissimus (Fr.) Fr.	644042	041160	1
L. obscuratus (Lasch) Fr.	050030	000274	1
L. pyrogalus (Bull.ex.Fr.) Fr.	641032	177067	4
L. quietus (Fr.) Fr.	757372	177373	13
L. rufus (Scop.ex.Fr.)Fr.	454212	177377	9
L. serifluus (DC.ex.Fr.) Fr.	557132	173375	8
L. subdulcis (Pers.ex.Fr.) Fr.	744172	037777	10
L. torminosus (Schaeff.ex.Fr.) Gray	004010	052366	2
L. turpis (Weinm.) Fr.	766172	177377	11
L. vietus (Fr.) Fr.	140052	003020	1
Russula aeruginosa Lindbl.ex.Fr.	445020	070204	1
R. alutacea (Pers.ex.Fr.) Fr.	074212	033015	2
R. atropurpurea (Krombh.) Britz.	776132	177377	11
R. brunneoviolacea Crawshay	144212	013043	2
R. cyanoxantha (Schaeff.ex.Secr.)Fr.	776172	177767	11
R. emetica (Schaeff.ex.Fr.) Gray	146132	157367	10
R. fellea (Fr.) Fr.	014412	024067	2
R. foetens (Pers.ex.Fr.) Fr.	454032	130061	2
R. fragilis (Pers.ex.Fr.) Fr.	744132	176375	6
R. grisea (Pers.ex.Secr.) Fr.	540032	167367	4
R. heterophylla (Fr.) Fr.	54132	017067	3
R. laurocerasi Melzer	003132	131126	3
R. luteotacta Rea	444030	131112	2
R. melliolens Quel.	500032	041041	1
R. ochroleuca (Pers.ex.Secr.) Fr.	777336	077777	20
R. rosea Quel.	614032	110175	3
R. sardonia Fr.	004032	056104	1
R. solaris Ferd. + Winge	000312	060021	1
R. sororia (Fr.) Rom.	051332	175367	6
R. vesca Fr.	540132	115006	2
R. xerampelina (Schaeff.ex.Secr.)Fr.	344032	146362	3
Clavariaceae (+ Cantharellaceae)			
Cantharellus cibarius Fr.	100032	021066	2
Clavulina cinerea (Bull.ex.Fr.) Schrot.	761562	041367	4
C. cristata (Holmsk.ex.Fr.) Schrot.	700132	163367	4
C. rugosa (Bull.ex.Fr.) Schrot.	700062	127364	3
Clavulinopsis helvola (Pers.ex.Fr.) Corner	654032	102177	3
C. luteoalba (Rea) Corner	040022	160200	1
Craterellus cornucopioides (L.ex. Fr.) Pers.	500020	131266	2

Pseudocraterellus sinuosus (Sow.ex. Fr.) Corn. + P. cinereus, P. pusillus	300120	011246	2
Ramaria stricta (Pers.ex.Fr.) Quel.	000606	020005	1
Hygrophoraceae			
Camarophyllus pratensis (Pers.ex. Fr.) Kumm.	000402	061227	1
C. virgineus (Wulf.ex.Fr.) Kumm. + C. niveus	020102	020062	1
Hygrocybe ceracea (Wulf.ex.Fr.) Kumm.	020002	020265	1
H. chlorophana (Fr.) Wunsche	000002	102064	1
Hygrocybe miniata (Fr.) Kumm.	000202	021265	1
H. psittacina (Schaeff.ex.Fr.) Wunsche	040402	121022	1
Hygrophorus hypothejus (Fr.ex.Fr.)Fr.	004202	011264	1
Tricholomataceae (+ Pleurotaceae)			
Armillaria mellea (Vahl.ex.Fr.) Kumm.	576377	172777	16
A. tabescens (Scop.ex.Fr.) Emel.	500130	040250	1
Asterophora lycoperdoides (Bull. ex.Mer.) Gray	610020	040127	1
A. parasitica (Bull.ex.Fr.) Sing.	310020	001062	1
Baeospora myosura (Fr.ex.Fr.) Sing.	004212	153305	2
Clitocybe brumalis (Fr.ex.Fr.) Quel.	560012	122260	1
C. cerrussata (Fr.) Gill.	440062	000022	1
C. clavipes (Pers.ex.Fr.) Karst.	576172	177377	9
C. dealbata (Sow.ex.Fr.) Karst.	563013	142264	3
C. ditopus (Fr.ex.Fr.) Gill.	050072	002235	1
C. flaccida (Sow.ex.Fr.) Karst. + C. inversa, C. vernicosa	652332	177266	6
C. fragrans (Sow.ex.Fr.) Karst.	471333	142276	4
C. fritilliformis (Lasch) Gill.	040122	052000	1
C. infundibuliformis (Schaeff.ex. Weinm.) Quel.	244132	177377	6
C. nebularis (Batsch ex.Fr.) Karst.	342012	166064	3
C. obsoleta (Batsch ex.Fr.) Quel.	563233	062226	2
C. odora (Bull.ex.Fr.) Karst.	560132	032347	3
C. rivulosa (Pers.ex.Fr.) Karst.	021033	122045	1
C. trullaeformis (Fr.) Karst. + C. forquignonii	100052	002100	1
C. vibecina (Fr.) Quel.	064312	004325	2
Collybia acervata (Fr.) Karst. + C. erythropus	576373	167367	11
C. butyracea (Bull.ex.Fr.) Karst.	776333	126367	5
C. cirrhata (Schum.ex.Fr.) Karst. + C. cookei, C. ocellata	546132	006060	3
C. confluens (Pers.ex.Fr.) Karst.	055012	013377	2
C. dryophila (Bull.ex.Fr.) Karst.	676177	177377	14

C. maculata (A+S.ex.Fr.) Karst.	054216	177777	9
C. succinea (Fr.) Quel.	044032	040340	1
C. tuberosa (Bull.ex.Fr.) Karst.	626232	072375	4
C. (Gymnopus) fusipes(Bull.ex.Fr.)Qu.	776572	173767	8
C. (Scorteus) fuscopurpurea (Pers. ex.Fr.) Karst.	052176	102343	4
C. (S) peronata (Bull.ex.Fr.) Karst.	776577	177377	15
C. (Tephrophana) asterospora (Lange) Ort.	010212	000006	1
Crepidotus mollis (Schaeff.ex.Fr.) Karst.	240022	126340	2
C. variabilis (Pers.ex.Fr.) Karst.	760162	177361	5
Crinipellis stipitarius (Fr.) Pat.	022001	102140	1
Hygrophoropsis aurantiacus (Wulf. ex.Fr.) Mre.	074373	156777	7
Laccaria laccata (Scop.ex.Fr.) B.+ Br. laccata	777373	177377	26
L.l. amethystina	777372	177367	15
L.l. bicolor + L.l. semiamethystina	255332	033266	4
L.l. rosella	657373	023377	8
Lyophyllum decastes (Fr.ex.Fr.)Sng.	140032	123240	1
Macrocystidia cucumis (Pers.ex.Fr.) Heim.	044442	040125	1
Marasmius androsaceus (L.ex.Fr.)Fr.	154012	173175	3
M. epiphyllus (Pers.ex.Fr.)Fr.	044442	040125	1
M. graminum (Lib.) Berk.	230030	014160	1
M. ramealis (Bull.ex.Fr.) Fr.	740562	075365	4
M. rotula (Scop.ex.Fr.) Fr.	062172	136340	3
M. (Scorteus) oreades (Bolt.ex. Fr.) Fr.	026413	172347	4
Melanoleuca excissa (Fr.) Sing.	121041	022055	1
M. melaleuca (Pers.ex.Fr.) Murr. + M. brevipes	301112	102245	1
Mycena alcalina (Fr.ex.Fr.) Karst.	776372	173377	12
M. amicta (Fr.) Quel.	106363	023353	3
M. avenacea (Fr.) Quel. (= M. olivaceomarginata + M. albolilacea, M. purpureofusca)	460402	162343	2
M. cinerella Karst.	000052	040321	1
M. epipterygia (Scop.ex.Fr.) Gray	664010	000036	1
M. epipterygioides Pears.	454000	142361	2
M. filopes (Bull.ex.Fr.) Karst.	776372	173377	16
M. flavoalba (Fr.) Quel.	763112	023375	4
Mycena galericulata (Scop.ex.Fr.) Karst.	776777	177777	18
M. galopus (Pers.ex.Fr.) Karst.	777777	177377	20
M. hemisphaerica Peck	240060	142240	1
M. inclimata (Fr.) Quel.	352032	123377	4
**M. laevigata (Lasch) Quel.	044012	012101	1
M. metata (Fr.ex.Fr.) Karst. + M. pumila	761073	163377	7
M. polygramma (Bull.ex.Fr.) Gray	762062	167376	6
M. pura (Pers.ex.Fr.) Karst.	542772	137367	6

M. sanguinolenta (A+S.ex.Fr.) Karst.	777777	177377	21
M. speirea (Fr.ex.Fr.) Gill.	340062	141752	2
M. stylobates (Pers.ex.Fr.) Karst.	664356	056767	4
M. vitilis (Fr.) Quel.	460022	142100	1
M. (Trogia) acicula (Schaeff.ex. Fr.) Karst.	040222	010152	1
M. (T.) gypsea (Fr.) Quel.	242472	062166	2
M. (Omphalina) fibula (Bull.ex.Fr.) Kuhn. + M. swartzii	764272	177377	7
Oudemansiella radicata (Relh.ex.Fr.) Sing.	124432	167240	2
Panellus stipticus (Bull.ex.Fr.) Karst.	040022	023640	1
Pleurotus cornucopiae (Paul.ex. Fr.) Roll.	020102	010555	1
P. dryinus (Pers.ex.Fr.) Karst.	060102	002217	1
P. ostreatus (Jacq.ex.Fr.) Karst.	060030	100334	1
Rhodotus palmatus (Bull.ex.Fr.) Mre.	064000	122220	1
Tricholoma imbricatum (Fr.ex.Fr.) Karst.	040012	120300	1
T. saponaceum (Fr.) Karst. + T. atrovirens, T. cnistum	100020	023204	1
T. sulphureum (Bull.ex.Fr.) Karst.	046032	120044	1
T. vaccinum (Pers.ex.Fr.) Karst.	104212	003177	2
T. virgatum (Fr.ex.Fr.) Karst. + T. hordum	141022	102005	1
Tricholomopsis platyphylla (Pers. ex.Fr.) Sing.	640062	034433	2
T. rutilans (Schaeff.ex.Fr.) Sing.	444212	071777	4
Rhodophyllaceae (+ Clitopilaceae)			
Entoloma rhodopolium (Fr.) Kumm.	042022	120321	1
Lepista nuda (Bull.ex.Fr.) Cke.	363122	130021	2
Leptonia euchroa (Pers.ex.Fr.) K.	200120	000163	1
*L. molliuscula Quel.	240100	030013	1
**L. omphaliformis Rom.	202042	021025	1
Nolanea cetrata (Fr.ex.Fr.) Karst.	544132	161365	3
N. mammosa (L.ex.Fr.) Quel.	004017	143245	2
N. papillata Bres.	046042	030340	1
N. sericea (Bull.ex.Mer.) Ort.	424202	023761	2
N. staurospora Bres.	640442	023375	4
Paxillopsis fallax (Quel.) Lange.	440242	003102	1
Cortinariaceae (+ Strophariaceae)			
Cortinarius cinnamomeus (Fr.) Fr.	034212	163027	2
C. croceoconus Fr.	004220	012025	1
C. leucopus Fr.	000022	120326	2
C. semisanguineus (Fr.) Gill.	014000	022031	1
C. tabularis (Bull.ex.Fr.) Fr.	400022	020222	1
C. (Telamonia s.l.) acutus (Pers. ex.Fr.) Fr.	241032	063325	3

C. (T.) hemitrichus (Pers.ex.Fr.) Fr.	144032	105040	1
C. (T.) holophaeus Henry	404022	041020	1
C. (T.) junghuhnii Fr.	504132	043320	3
C. (T.) paleaceus Fr. + C. flexipes	441012	002373	3
C. (T.) uraceus Fr.	740132	163367	5
Crepidotus: see Tricholomataceae			
Deconica crobulus (Fr.) Rom.	040362	000343	1
D. montana (Pers.ex.Fr.) Ort.	004043	132364	3
D. physaloides (Bull.ex.Mer.) Karst.	040012	003064	1
Galerina graminea (Vel.) Kuhn.	024202	100016	1
G. hypnorum (Schrank ex.Fr.) Kuhn.	677373	177377	10
G. mycenopsis (Fr.ex.Fr.) Kuhn.	104252	173376	4
G. sideroides (Bull.ex.Fr.) Kuhn. + G. ampullaceocystis	474672	112337	3
G. (Kuhneromyces) marginatus (Secr.) Sing.	044002	140220	1
G.(K.) mutabilis (Schaeff.ex. Fr.) Sing.	341252	002275	2
G.(K.) unicolor (Vahl.ex.Sommerf.) Sing.	240410	060320	1
Gymnopilus penetrans (Fr.ex.Fr.) Murr. + G. liquiritiae	004212	010303	2
G. sapineus (Fr.) Murr.	004312	112176	2
*G. spectabilis (Fr.ex. Kumm.)Sing.	024022	060305	1
Hebeloma crustuliniforme (Bull.ex. St.Am.) Quel.	505132	173067	4
H. mesophaeum (Pers.) Quel. + H. testaceum	175272	077767	5
H. sacchariolens Quel.	620072	140143	2
Hypholoma fasciculare (Huds.ex. Fr.)Karst.	776773	177377	23
H. sublateritium (Fr.) Quel.	564062	172364	3
Inocybe abjecta (Karst.) Sacc.	100112	000361	1
I. asterospora Quel.	755262	057367	4
I. deglubens (Fr.) Fr. + I. flocculosa	200030	110166	1
I. fastigiata (Schaeff. ex. Fr.) Quel.	140032	062241	1
I. geophylla (Sow.ex.Fr.) Karst.	704120	123344	1
I. lacera (Fr.) Fr.	214020	110005	1
I. napipes Lange	340002	141044	1
I. petiginosa (Fr.ex.Fr.) Gill.	440202	020044	1
Naucoria escharoides (Fr.ex.Fr.) Karst.	442242	173576	5
N. acolecina (Fr.) Quel.	040242	011656	1
Phaeomarasmius erinaceus (Fr.) Kuhn.	240122	062241	2
Pholiota inaurata (WGSm.) Moser	024122	022044	1

Psilocybe semilanceata (Fr.ex.Secr.) Karst.	667073	142365	4
Stropharia aeruginosa (Curt.ex.Fr.) Quel.	562276	162377	7
S. coronilla (Bull.ex.Fr.) Quel.	060042	163100	1
S. inuncta (Fr.) Quel.	022052	102244	1
S. semiglobata (Batsch ex.Fr.) Quel.	170243	000371	2
Tubaria furfuracea (Pers.ex.Fr.) Gill. + T. conspersa	574176	067367	5
Coprinaceae (+ Bolbitiaceae)			
*Agrocybe pediades (Fr.) Fay.	024000	060003	1
Bolbitius vitellinus (Pers.ex.Fr.) Fr.	022633	166375	3
Conocybe laricina (Kuhn.) Kuhn.	060642	040102	1
C. tenera (Schaerf.ex.Fr.) Kuhn. + C. microspora, C. spicula, C. subovalis	776773	167375	10
C. (Pholiotina) blattaria (Fr.) Kuhn. + C. filaris	041202	040007	1
C. (P.) togularis (Bull.ex.Fr.) Kuhn.	002660	102142	1
Coprinus atramentarius (Bull.ex. Fr.) Fr.	760032	172775	4
C. comatus (Mull.ex.Fr.) Gray	520230	133155	2
C. disseminatus (Pers.ex.Fr.) Gray	162052	041327	3
C. hemerobius Fr.	260322	123367	4
C. lagopus (Fr.) Fr. + C. lagopides	562672	167337	4
C. micaceus (Bull.ex.Fr.)	776572	156367	9
C. plicatilis (Curt.ex.Fr.) Fr.	660672	177773	8
Lacrimaria velutina (Pers.ex.Fr.) K.+M.	760162	137372	4
Panaeolus acuminatus (Schaeff.ex. Secr.) Quel. + P. rickenii	070403	034305	3
P. campanulatus (Bull.ex.Fr.) Quel.	401052	170163	3
P. sphinctrinus (Fr.) Quel.	462400	104141	1
P. subbalteatus (Berk. + Br.) Sacc.	020601	152010	1
Psathyrella candolleana (Fr.) Mre.	460222	025744	2
P. caudata (Fr.ex.Fr.) Quel.	462213	107277	2
**P. fibrillosa (Pers.ex.Fr.) Pears. + Denn.	365152	132760	3
P. gracilis (Fr.) Quel.	572672	173767	7
P. hydrophila (Bull.ex.Mer.) Mre.	761172	173377	8
P. microrhiza (Lange) Konr. + Maubl. + P. pseudobifrons	440002	021242	1
P. obtusata (Fr.) ABSm.	566042	176373	5
P. pennata (Fr.) Pears. + Denn. + P. carbonicola, P. ochrospora	340032	133367	3

Agaricaceae s.s.			
Agaricus bisporus (Lange) Pil.	260403	130301	1
A. campestris L.ex.Fr. + A. lividonitidus	022002	152001	1
A. impudicus Rea + A. variegans	002003	100247	1
A. semotus Fr. + A. purpurellus	020032	070100	1
A. sylvaticus Schaeff.ex.Secr.	263353	072367	4
A. sylvicola (Vitt.) Peck	463522	136327	3
A. xanthodermus Genev.	020000	020144	1
Cystoderma amianthinum (Scop.ex. Fr.) Fay.	030252	002266	2
C. granulosum (Batsch ex.Fr.) Fay.	105012	062223	1
Lepiota excoriata (Schaeff.ex.Fr.) Karst.	440012	141100	1
L. procera (Scop.ex.Fr.) Gray	275132	123367	5
L. rhacodes (Vitt.) Quel.	666332	062767	7
L. (Lepiotula) clypeolaria (Bull. ex.Fr.) Karst.	020412	100100	1
L. (L.) cristata(Fr.) Karst.	060010	102140	1
L. (L.) felina (Pers.ex.Fr.) Karst.	056316	053767	5
L. (L.) fuscovinacea Mǿll.+ Lange	042002	000320	1
L. (Leucoagaricus) leucothites (Vitt.) Ort. + L. holosericea	064003	147766	2
Volvariaceae			
Pluteus cervinus (Schaeff.ex.Fr.) Karst.	762232	177775	9
P. salicinus (Pers.ex.Fr.) Karst.	742772	163777	7
P. thomsonii (B.+Br.) Denn. + P. cinereus, P. cinereofuscus, P. godeyi	040242	003461	1
Volvariella parvula (Weinm.) Ort.	420010	100062	1
Amanitaceae			
Amanita citrina (Schaeff.) Gray	557172	177377	13
A. excelsa (Fr.) Karst.	601120	031065	2
A. fulva (Schaeff.) Secr.	045232	177367	8
A. gemmata (Fr.) Gilb. var. exannulata	004202	140006	1
A. muscaria (L.ex.Fr.) Hooker	544172	167777	8
A. phalloides (Vaill.ex.Fr.) Secr.	152012	173377	4
A. rubescens (Pers.ex.Fr.) Gilb.	777377	175777	12
A. vaginata (Bull.ex.Fr.) Vitt.	457112	075366	3
Boletineae			
Boletus edulis Bull.ex.Fr.	106112	171162	3
B. pinicola Venturi	014012	001346	1
B. erythropus (Fr.ex.Fr.) Secr.	140032	030206	1

Gomphidius glutinosus (Schaeff.ex. Fr.) Fr.	004002	030063	1
G. roseus (Fr.) Karst.	004012	032632	2
G. rutilus (Schaeff.ex.Fr.) Lund.	020002	110340	1
Leccinum scabrum (Bull.ex.Fr.) Gray	744032	173374	7
*L. aurantiacum (Bull.) Gray	244022	012164	1
Paxillus involutus (Batsch ex. Fr.) Fr.	777773	177737	17
Plicaturella atrotomentosa (Batsch ex.Fr.) Murr.	004112	001767	4
P. panuoides (Fr.ex.Fr.) Murr.	040112	002162	1
Suillus bovinus (L.ex.Fr.)OKze.	024012	003277	2
S. elegans (Schum.ex.Fr.) Snell	004002	103300	1
S. granulatus (L.ex.Fr.) OKze.	004012	011266	2
S. luteus (L.ex.Fr.) Gray	005212	123377	6
S. piperatus (Bull.ex.Fr.) OKze.	554012	173167	4
S. variegatus (Sow.ex.Fr.) OKze.	014200	031027	1
Xerocomus badius (Fr.) Gilb.	755072	117277	6
X. chrysenteron (Bull.ex.StAm.)Quel.	766772	177777	13
X. leoninus (Pers.) Quel.	041020	011406	1
X. parasiticus (Bull.ex.Fr.) Quel.	040032	153151	2
X. subtomentosus (L.ex.Fr.) Quel.	456732	133277	6

LYCOPERDALES

Bovista nigrescens Pers.	021010	112001	1
B. plumbea Fr.	030001	110322	1
Calvatia depressa (Bon.) Vitt.	424011	132207	2
C. bovista (L.ex.Fr.) Fr.	12001	040200	1
C. excipuliforme (Scop.ex.Pers.) Fr.	041012	021377	2
C. gigantea (Batsch ex.Pers.) Fr.	064100	020345	1
Lycoperdon atropurpureum Vitt.	414362	152356	3
L. echinatum Pers.	204302	000165	2
L. ericetorum Pers.	446031	177346	3
L. perlatum Pers.	776337	177363	12
L. pyriforme (Schaeff.) Pers.	743776	173777	6
Mutinus caninus (Huds.) Fr.	142272	071365	3
Phallus impudicus (L.) Pers.	777377	177777	19
Scleroderma aurantium Pers.	545131	137777	11
S. verrucosum (Vaill.) Pers.	052312	027337	4

Note on nomenclature: for the Agaricales I have used the names given in Dennis Ortin and Hora's Check List [8], except in the Hygrophoraceae and Boletineae where I conform to the general modern practice in regard to generic names; a few species misnamed in the Check List, marked * above, have been corrected. There are further a few Agaricales in the above list which do not appear in the Check List, and these are marked **.

It will be noted that several common species are only poorly represented, or fail to appear, in the above list; for example, in the genus *Agaricus*. This is in most cases due to the observations being heavily biased in favour of forest communities. On the other hand, a few species usually considered rare are included: these are for example *Phellinus torulosus* (where we include the first record of this species in Gt. Britain) and *Xerocomus leoninus*, which appear to be locally not uncommon. The same probably applies to *Mycena laevigata* which is not in Dennis Ortin and Hora's Check List; but *Leptonia omphaliformis* may be quite generally distributed in Britain. *Psathyrella fibrillosa*, also omitted from the Check List, is here understood in the sense of Lange; it is not uncommon.

APPENDIX II

List of Localities Examined

Digit 1 Bit 1 : *Assington Thicks* : Partly self-sown forest with much Birch, dating from 1917, and partly young Pine plantation. Oak and Ash present. (158 spp.)

Digit 1 Bit 2 : *Bentley Long Wood* : Oak-Ash wood on chalky boulder clay; owing to narrow shape the typical forest microclimate may not everywhere develop. (120 spp.)

Digit 1 Bit 3 : *Chalkney Wood* near Earles Colne, Essex : Mainly Oak and Chestnut on hilly ground, with some Alder in the bottoms. Felled in 1963 and replanted with Pine. (139 spp.)

Digit 2 Bit 1 : *Dodnash Wood* : Rather open Chestnut plantation with many other deciduous species present, including much Alder. Part forms a young Pine plantation much of the open ground is covered in a dense *Pteridium* and *Rubus* brakes. (234 spp.)

Digit 2 Bit 2 : This refers to the area in the immediate vicinity of *Flatford Mill* : pasture land, gardens, hedgerows, etc., together with some estuarine flats. (143 spp.)

Digit 2 Bit 3 : *Gobblecock* new Hollesley : Heathland with some old Oak. (87 spp.)

Digit 3 Bit 1 : *Hollesley Heath* : Heathland with much self-sown Pine and some coniferous plantation (age about 20 years in 1952). (149 spp.)

Digit 3 Bit 2 : *Fishpond Wood*, East Bergholt : A small Oak-Ash wood on alluvial soil at the edge of the flood-plain on the Stour. (107 spp.)

Digit 3 Bit 3 : *Great Martins Wood*, East Bergholt: A variety of plantations, Pine, Ash, etc., on gravelly soil. (59 spp.)

Digit 4 Bit 1 : *Mistley Wood*, Essex : Oak, Ash and Beech on steep slope overlooking the Stour estuary; also some grassland. (51 spp.)

Digit 4 Bit 2 : *Pin Mill Woods* : Oak, Ash and Beech on slopes overlooking the Orwell estuary, together with much planted Pine. (116 spp.)

Digit 4 Bit 3 : *Copperas Wood*, near Wrabness, Essex : Chestnut coppice with some old Oak-Ash; mainly similar to Stour Wood. (124 spp.)

Digit 5 Bit 1 : *Arger Fen* near Assington : not a fen, but a hilly wood consisting mainly of rather open Oak and Ash. (116 spp.)

Digit 5 Bit 2 : *Stour Wood* near Wrabness, Essex; a commercial Chestnut plantation managed as coppice; a few other trees present. (198 spp.)

Digit 5 Bit 3 : *Staverton Thicks* : comprises (a) primary Oak-Holly forest of open structure with much bracken, (b) a number of old Beech trees planted by a roadside, (c) a mature Pine plantation, and (d) a young Pine plantation. (215 spp.)

Digit 6 Bit 1 : *Shotley Wood* : Oak-Ash wood on low-lying clayey soil. (22 spp.)

Digit 6 Bit 2 : *Commons Wood*, East Bergholt : Ornamental woodland and gardens, with some Pine plantations; large variety of vegetation. (285 spp.)

Digit 6 Bit 3 : *Shingle Street* : Littoral grassland and hedgerows, with shingle foreshore. (49 spp.)

APPENDIX III

Composition of Super-Classes

List 1: Superclasses obtained from OP/*L* using cohesion G_1 including all localities

Superclass A (mainly species of deciduous woodlands)

Lactarius blennius	L. decipiens	Craterellus cornucopioides
Pseudocraterellus sinuosus	Asterophora lycoperdoides	
A. parasitica	Pleurotus cornucopiae	Leptonia euchroa
L. molliuscula	Camarophyllus pratensis	
C. virgineus	Hygrocybe ceracea	H. chlorophana
Inocybe deglubens	Lactarius piperatus	Tricholoma saponaceum

Superclass B (ecological character uncertain)

Lactarius helvus	Macrocystidia cucumis	Mycena epiptergia
M. cinerella	Leptonia omphaliformis	
Nolanea papillata	Naucoria scolecina	Deconica montana
Pluteus thomsonii	Grifola frondosa	Polyporellus varius

Superclass C (including species of grassland and deciduous woods)

Lactarius vellereus	Russula laurocerasi	Cantharellus cibarius
Clitocybe rivulosa	Cortinarius leucopus	Agaricus semotus
Russula drimeia	R. solaris	

Superclass D (species of coniferous woods, birch, and heath)

Lactarius blennius	L. decipiens	L. deliciosus
L. helvus	L. piperatus	L. tomentosus
Russula drimeia	R. fellea	R. solaris
Ramaria stricta	Camarophyllus pratensis	
Hygrocybe miniata	H. ceracea	H. chlorophana
Hygrophorus hypothejus	Collybia maculata	
Clitocybe trullaeformis		Tephrophana asterospora
Mycena cinerella	Tricholoma saponaceum	T. vaccinum
Nolanea mammosa	Deconica montana	Cortinarius cinnamomeus

Superclass D (contd.)

C. semisanguineus	Galerina mycenopsis	Gymnopilus penetrans
G. sapineus	Inocybe abjecta	Amanita gemmata
Cystoderma amianthinum	Boletus pinicola	
Gomphidius roseus	G. glutinosus	Plicaturella panuoides
Suillus elegans	S. granulatus	S. variegatus
Auriscalpium vulgare	Calcera viscosa	Coltricia perennis
Hirschioporus abietinus		Tyromyces chioneus
Thelephora intybacea		

<u>Superclass E</u> (Grassland and littoral species)

Agaricus campestris	A. impudicus	A. xanthodermus
Camarophyllus pratennis		C. virgineus
Hygrocybe ceracea	H. chlorophana	Melanoleuca excissa
Crinipellis stipitarius		Pleurotus cornucopiae
Gomphidius rutilus	Bovista plumbea	Calvatia bovista

<u>Superclass F</u> (species associated with heath vegitation)

Camorophyllus pratensis		Hygrocybe ceracea
H. chlorophana	Agrocybe pediades	Cortinarius semisanguineus
Coltricia perennis		
Gomphidius glutinosus	G. rutilus	Suillus elegans

<u>List 2</u>: Superclasses obtained from OP/*L* using cohesion G_1 with three diversified localities omitted.

<u>Superclass G</u> (species of deciduous woods)

Lactarius camphoratus	L. decipiens	L. obscuratus
L. vellereus	L. piperatus	Russula solaris
Cantharellus cibarius	Clavulinopsis luteoalbus	
Clitocybe fritilliformis		
Mycena acicula	Panellus stipticus	Lyophyllum decastes
Tricholoma saponaceum	T. virgatum	Leptonia molliuscula
Entoloma rhodopolium	Cortinarius leucopus	Inocybe abjecta
I. fastigiata	Agaricus semotus	Boletus erythropus
Plicaturella panuoides	Pleurotus dryinus	Xerocomus leoninus
X. parasiticus	Phlebia radiata	

Superclass H (further species of deciduous woods)

Lactarius blennius	L. decipiens	L. obscuratus
Russula foetens	R. rosea	Clavulinopsis helvola
Asterophora lycoperdoides		Clitocybe ditopus
Marasmius graminum	Inocybe lacera	

Superclass I (species of grassland and deciduous woods)

Lactarius decipiens	Russula solaris	Camarophyllus pratensis
C. virgineus	Hygrocybe psittacina	Pleurotus cor-nucopiae
Leptonia molliuscula	Galerina unicolor	Plicaturella panuoides
Bjerkandera fumosa		

Superclass J (species of grassland and heaths)

Crinipellis stipitarius		Panaeolus acumi-natus
P. subbalteatus	Agaricus bisporus	Lepiota leucoth-ites
Bovista plumbea	Calvatia bovista	

Superclass K (species of coniferous or mixed woods)

Lactarius deliciosus	L. helvus	L. hepaticus
L. rufus	L. tomentosus	L. vietus
Russula alutacea	R. brunneoviolacea	R. fellen
Ramaria stricta	Hygrocybe miniata	Hygrophorus hypothejus
Collybia maculata	Clitocybe trullaeformis	
C. vibecina	Collybia confluens	
Baeospora myosura	Tephrophana asterospora	
Macrodystidia cucumis	Marasmius androsaceus	M. epiphyllus
Mycena cinerella	M. epipterygioides	M. laevigata
Tricholoma vaccinum		Tricholomopsis rutilans
Nolanea mammosa	N. papillata	Deconica montana
Cortinarius cinnamomeus		C. paleaceus
C. semisanguineus	Galerina graminea	G. mycenopsis
Galerina marginatus	Gymnopilus penetrans	
G. sapineus	Inocybe petiginosa	Nolanea scolecina
Agrocybe pediades	Conocybe blattaria	Psathyrella microrhiza

Superclass K (contd.)

Agaricus impudicus	Amanita gemmata	Lepiota excoriata
L. fuscovinacea	Pluteus thomsonii	Boletus pinicola
Gomphidius glutinosus	G. roseus	Suillus elegans
S. granulatus	S. luteus	S. variegatus
Auriscalpium vulgare	Coltricia perennis	Grifola frondosa
Polyporellus varius	Thelephora terrestris	T. intybacea

List 3: Superclasses obtained from OP/*L* using cohesion G_2 with the three diversified localities omitted.

Superclass L (species of deciduous woods)

Russula adusta	R. rosea	Clavulinopsis helvola
Craterellus cornucopioides		Armillaria tabescens
Asterophora parasitica	Clitocybe clavipes	C. infundibuliformis
C. odora	Collybia acervata	Mycena inclinata
Galerina hypnorum	Coprinus comatus	Tyromyces lacteus
Clitocybe flaccida	Hypholoma fasciculare	Stereum rugosum

Superclass M (species of deciduous woods)

Russula rosea	Marasmius graminum	Mycena hemisphaerica
M. inclinata	M. polygramma	Omphalina fibula
Lepista nuda	Cortinarius uraceus	Hebeloma mesophaeum
Inocybe asterospora	Phaeomarasmius erinaceus	
Concybe tenera	Coprinus hemerobius	Lepiota procera
Heteroporus biennis		

Superclass N (species of deciduous woods)

Lactarius piperatus	Russula ochroleuca	R. vesca
Cantharellus cibarius	Clavulina cinerea	C. rugosa
Armillaria tabescens	Asterophora parasitica	
Clitocybe clavipes	C. flaccida	C. infundibuliformis
Amanita excelsa	Lycoperdon pyriforme	

Superclass O (species of deciduous woods)

Russula atropurpurea	R. vesca	Clitocybe obsoleta
Mycena pura	Omphalina fibula	Tricholomopsis platyphylla
Deconica crobulus	Hypholoma sublateritium	
Psathyrella candolleana		Agaricus bisporus
Bovista plumbea	Calvatia bovista	C. depressa
Ionotus hispidus		

Superclass P (mainly from deciduous woods)

Paxillopsis fallax	Deconica crobulus	Stropharia aeruginosa
S. semiglobata	Galerina sideroides	Inocybe asterospora
Naucoria escharoides	Conocybe laricina	C. tenera
Psathyrella candolleana		Amanita rubescens
Cystoderma amianthinum	C. granulosum	
Pluteus salicinus	Scleroderma aurantium	

Superclass Q (deciduous and mixed woods)

Lactarius pyrogalus	Russula laurocerasi	Clavulina cinerea
C. cristata	C. rugosa	Cratarellus cornucopioides
Pseudocraterellus sinuosus	Collybia cirrhata	
Mycena polygamma	Omphalina fibula	Cortinarius acutus
Hebeloma crustuliniforme		

Superclass R (deciduous and mixed woods)

Russula nigrescens	Asterophora parasitica	Clitocybe flaccida
C. nebularis	Marasmius graminum	Mycena hemisphaerica
M. inclinata	M. polygamma	M. speirea
Tricholomopsis platyphylla		Inocybe lacera
Lacrimaria velutina		

Superclass S (mainly from deciduous woods)

Lactarius mitissimus	L. pyrogalus	L. subdulcis
L. turpis	Russula heterophylla	R. luteotacta
R. vesca	Clavulina cinerea	Mycena epipterygia
M. filopes	Hebeloma crustuliniforme	
H. sacchariolens		

Superclass T (deciduous woods, especially with beech)

Russula grisea	R. melliolens	R. rosea
R. vesca	R. xerampelina	Cantharellus cibarius
Clavulinopsis helvola	Clitocybe clavipes	C. odora
Collybia fusipes	C. peronata	Dacrymyces deliquescens

Superclass U (deciduous woods, especially with oak)

Clitocybe obsoleta	Collybia acervata	C. butyracea
C. cirrhata	C. dryophila	C. tuberosa
C. fusipes	C. peronata	Mycena inclinata
	Omphalina fibula	Entoloma rhodopolium
Galerina unicolor	Lycoperdon pyriforme	Hapalopilus nidulans
Inonotus hispidus	Polyporus squamosus	Vuilleminia comedens

Superclass V (species of coniferous woods)

Russula brunneo-violacea	Clitocybe vibecina	
Collybia succinea	Tricholomopsis rutilans	
Lepiota felina	Amanita fulva	Suillus luteus
S. variegatus	Lycoperdon echinatum	Scleroderma verrucosum
Auriscalpium vulgare	Calocera viscosa	Fomitopsis annosus
Tyromyces chioneus		

Superclass W (species associated with pine, birch, and heath)

Russula heterophylla	R. rosea	Marasmius andro-saceus
Mycena epipterygioides	Nolanea sericea	
Stropharia semiglobata	Cortinarius cinnamomeus	
Galerina sideroides	Tubaria furfuracea	Lepiota clypeol-aria
Suillus piperatus	Lycoperdon atropurpureum	
Piptoporus betulinus		

Superclass X (species of birch and pine)

Lycoperdon hepaticum	Russula brunneoviolacea	
R. xerampelina	Collybia maculata	Marasmius androsaceus

Superclass X (contd.)

Cortinarius hemitrichus	C. semisanguineus	
Galerina mycenopsis	Hebeloma mesophaeum	Inocybe astero-spora
Suillus badius	S. piperatus	

Superclass Y (mainly species of coniferous woods)

Clitocybe vibecina	C. metachroa	Collybia tuberosa
Hygrophoropsis aurantiacus	Laccaria laccata bicolor	
Mycena stylobates	Cortinarius cinnamomeus	
C. croceoconus	Galerina graminea	G. sideroides
Gymnopilus sapineus	Xerocomus subtomentosus	
Lycoperdon perlatum		

Superclass Z (grassland fungi)

Crinipellis stipitarius		Marasmius oreades
Mycena avenacea	Bolbitius vitellinus	Panaeolus acuminatus
P. sphinctrinus	P. subbalteatus	Psathyrella caudata
Agaricus bisporus	Bocista plumbea	Calvatia bovista
C. depressa	Ionotus hispidus	

Superclass AE (mainly species of grassland)

Russula ochroleuca	Nolanea sericea	Deconica montana
Psilocybe semilanceata	Stropharia semiglobata	
Galerina hypnorum	G. sideroides	Coprinus micaceus
C. plicatilis	Psathyrella gracilis	Pluteus cervinus
Calvatia depressa		

Note on the characterizations of Superclasses: These are derived from the intersections of ecological descriptors pertaining to all the localities represented among the classes forming the Superclasses.

CLASSIFICATION OF MIXED MODE DATA BY R-MODE FACTOR ANALYSIS AND Q-MODE CLUSTER ANALYSIS ON DISTANCE FUNCTION

James M. Parks

Lehigh University, Bethlehem, Pa., U.S.A.

Classification in the broad sense, and numerical taxonomy in a somewhat more narrow sense, is the procedure of putting similar objects together into groups and dissimilar objects into different groups, with the number of groups and the descriptive or discriminant characteristics of the groups unknown. The number of variables measured on each object should be large, and ideally the variables should be uncorrelated, *i.e.*, not measuring the same fundamental attribute. Many different measures of similarity have been used, and several procedures for clustering have been described, but none appear to be ideal for all situations. The purpose of this paper is to describe a computer classification system which appears to have some features approaching universal application. In order to handle the types of data sets encountered in numerical taxonomy and other classification problems, I set a minimum capability of 200 variables and 1000 samples as the desired goal in writing a new computer program.

Most of the readily available computer programs for numerical taxonomy (or for the more general problem of objective classification) are deficient in one or more of the following aspects:

1. Limitation in number of variables or number of samples (or their combination) that can be handled;

2. Lack of method for eliminating or otherwise adequately manipulating redundant variables;

3. Inability to handle mixed mode data (continuously variable measurements, integer counts, presence or absence, and coded states);

4. Final dendrogram of results must be drawn by an X-Y plotter, or by hand.

The mixed mode problem is perhaps the most difficult to solve. Of the many measures of similarity that have been proposed and used, the distance function appears to come closest to being able to handle mixed mode data. The form of distance function used in this paper uses normalized (transformed to range from 0.0 to 1.0) values of the variables. The sum of the squared differences is divided by the number of variables used, so that the resulting distance ranges from 0.0 (exact similarity) to 1.0 (complete dissimilarity). Missing data can be handled by changing this divisor of number of variables used. The formula is:

$$D_{i,j} = [\sum_{k=1}^{M} (X_{ki} - X_{kj})^2/M]^{\frac{1}{2}} \qquad (1)$$

With this form of distance function, presence-absence data can be coded 0 = absence, 1 = presence, and missing data can be coded -1. Multistate coded data can similarly use -1 for missing data, and the several states will be normalized so that the largest code number used will be transformed to 1.0 and intermediate states to evenly spaced fractional numbers. The negative 1 code can be used for missing data in integer count and continuously variable measurement data as long as negative measurements are not allowed. Normalizing these data will prevent inadvertent "weighting" when comparing a variable with large values to one with small values.

The principal difficulty with the distance function as a measure of similarity is that it is a Euclidean distance, which assumes that the variables are orthogonal or uncorrelated. In most classification problems, as many variables as possible are used, and inevitably some are redundant or highly correlated, and are thus not orthogonal. Many of the other measures of similarity are unable to take this into account and thus some fundamental attributes, such as size, are heavily weighted in the classification scheme because many of the original variables are highly correlated with a size factor.

Before the Q-mode cluster analysis is performed, it is therefore necessary to transform the variables to an uncorrelated or orthogonal set of variables. This is done by performing an R-mode factor or principal components analysis on the normalized variables, and then estimating factor scores or factor measurements for each sample. The usual R-mode factor analysis is performed on a matrix of product-moment correlation coefficients, but these are inadequate for mixed-mode data. Consequently an R-mode matrix of normalized distance function coefficients is used for the starting point of an R-mode factor analysis.

Most of the available factor analysis or eigenvalue-eigenfactor computer programs are limited by the number of variables they can handle. I have written a factor analysis subroutine based upon an older iterative successive squaring of the matrix method, which although slightly slower than the usual Jacobi eigenvalue routines, does not require the whole matrix to be in memory at once, and can therefore be expanded to handle 200 variables on a medium size computer.

Generally, the number of significant factors will be less than the original number of variables, due to the redundant and correlated nature of the original data set. For the clustering subroutine, I have set a limit of 20 transformed variables. Provision is made for using up to 20 factors in estimating the factor measurements, with cut-offs based on a pre-set percentage of the variance to be accounted for, or a minimum percentage of the variance to be accounted for by the last factor used. When 85-90% of the variance is accounted for, there is little significant loss of information in the resulting classification. It is not necessary to rotate the factors to a simple solution before estimating the factor measurements, as the distances will remain the same for rotated or unrotated factors.

For the clustering subroutine, the factor measurements are normalized, and the distance functions between all possible pairs of samples are calculated. Some economy of computer time is achieved by performing the clustering on the squared mean differences, and only taking the square roots once at the end of the procedure. The entire matrix of Q-mode distance coefficients is not stored in memory at one time: in order to achieve the goal of being able to handle 1000 samples a method was developed in which only a limited number of the smallest distance functions are stored at any one time, and all the distance functions are recomputed at each iteration. A modification of Sokal and Michener's pair-group, unweighted average linkage method of clustering is used. Several pairs may be utilized on each iteration as long as none of the pairs interact, giving the effect of a variable-group method.

The hierarchical diagram of relationships (dendrogram) is plotted with distances along the X-axis, and alphanumeric sample designators along the Y-axis. This is conveniently done by a high speed printer, preferably off-line from the computer. Small differences in distances between samples or groups are relatively unimportant, so the incremental plotting by printer is only a slight disadvantage. For the X-axis, 100 printer spaces are used, and the scale is expanded to spread the diagram out by setting the range of the scale to the largest distance encountered, which is rarely 1.0.

The program herein described appears to meet the desired criteria set out in the beginning of this paper. The program, in FORTRAN IV, is now operational on a CDC-6400, but has not yet been adequately tested on a variety of problems. The program will be published in the Kansas Computer Contribution Series, and copies of the source deck with adequate documentation are available from the author.

Due to difficulties with a remote terminal system during the change to the CDC-6400 computer this summer at Lehigh University, the multiple discriminant analysis program to test and adjust for optimum intra-group homogeneity and maximum inter-group separation is not yet operational.

Discussion

Q. (Sneath) How do you centre the stems on the dendrogram print-out?

A. I keep track of the number of O.T.U's in each cluster and space the stem half this number of lines from the first O.T.U. of each cluster.

Q. (Tomassone) As a statistician I am very much surprised to see that you have never used Mahalanobis' distance (specially for continuous variables) which have "good" properties. I only see one reason for this omission when you have no within-population dispersion matrix, but as soon as you have within variation, I think it is necessary to use it.

A. When I first started writing this program I did consider working with this distance function but I forget now the reasons why I didn't use it. I found that a simpler distance function seemed to be adequate for the amount of data I was using.

Q. (Saksena) May I ask which of the two R and Q techniques you are using in your cluster analysis?

A. I am doing the principal components analysis in R mode and the cluster analysis in Q mode.

Comment: (Saksena) The Mahalonobis' general distance function, I think, could be made use of if cluster analysis is done with the R technique as that would allow one to take into consideration the within sample variations, if any.

A. I'll look at it again. I have only recently been using prime components in R mode as an initial step.

Q. (Wishart) Concerning the standardisation or normalisation of factor scores vectors obtained from a principal component solution, it is the case that if those components having small variations are transformed then the resultant emphasis on these components is overrated by comparison with the major components and the resultant distance measures bear no relation to the original distances in standardised space. Thus, could you say what criterion you adopt when you choose those components which are to be transformed?

A. I do this largely on an empirical basis and I have not yet run enough different types of such data to see what the differences can be but I don't standardise, I normalise. This is only to give me the way of setting the maximum distance between any two samples down to one.

Comment: (Hope) This question of what components to retain and whether or not to standardise or normalise them is rather similar to the question of whether or not to rotate factors in factor analysis in the sense of there being very good theoretical reasons for rotating and very little practical guidance on how to rotate. Similarly for standardising components, I have done empirical investigations of this. For example with two samples I took principal components for these samples for five variables. I took the five principal components and compared the corresponding components over the two samples and I found that the components corresponded very closely with the cosine of .95 between the first component of one sample and the first component of the next sample and so on down to the smallest component which accounts for nearly 5% of the variance and these weren't particularly large samples. And there was also the theoretical consideration that the component space of the space of the first of the components accounting for the first 80% of variance may completely exclude the canonical space - the discriminant space. Theoretically if you take the two group case it may be that the Fisher's two group discriminant function may lie in the space of "umpteenth" component and you miss it out when you take only the first few components and this consideration generalises to the case where you have more than two groups in the canonical space. I think there are cases for missing out say the first component which is a size component because you can usually identify it and there are cases if you are retaining only the first three or four components for standardising those. These are theoretical cases because the weighting actually given to the components just simply depends upon what particular set of variables are chosen - I mean, the fact that you have chosen ten variables of a certain sort means that the first component measures that particular dimension. You might want to reduce those 10 variables to just 1 variable by just standardising the component but it would be a mistake to take 20 components, standardise them and use them all in those circumstances.

Comment: (Parks) In this general field I would like to make a plea for some standard sets of data about which we know a lot of answers which then could be used on various old and new methods so that we can see just what some of these things actually do in practice. It is all very well sometimes to talk in theory of what we should or shouldn't do but largely it depends on what kind of data we use and so several sets of standard data to be compared would be very useful.

Q. (Goronzy) Kruskal has suggested a method for metricising of binary data. The work is reported in Psychometrika. The approach would be valuable to deal with mixed mode data.

A. I am not aware of this but I will look for it.

Q. (Parker-Rhodes) You said you had a lot of trouble with getting a programme which would handle the amount of the data you had. How long would it take for you to finally get it working?

A. I have run it on six different computers for all different speeds. The original programme in "Mad" language on an old IBM 704 could run as long as three hours for 200 samples. The C.D.C. 6400 now is running over the same 200 sample problem in about 200 seconds. I think, I could speed this up.

Q. How long would you reckon it takes on the 1620?

A. It was a couple of years ago that I met the same problem. I would expect something of the order of 20 minutes and may be less.

Q. (Parker-Rhodes) Reverting to the question of normalising of data, with a lot of biological data one has trouble if you just take the largest and smallest. These are rather random and the things tend to cluster in rather a small space not very near the middle. Would you not find it better to take quantities or something for extreme points and not distinguish ones beyond.

A. This is quite possible. In the particular kind of data I have been using there haven't been too many of these wildly extreme points but if these did occur then it just could be a problem and there should be some way of overcoming it.

CLASSIFICATION TECHNIQUES FOR LARGE SETS OF DATA

G. J. S. Ross

Statistics Department, Rothamsted Experimental Station

Introduction

I wish to discuss first why it is necessary to be able to classify large sets of data, then to describe the classification techniques used at Rothamsted, and finally to show how these techniques can be extended to handle large sets of data without unreasonable demands on computer storage and time.

Characteristics of Large Sets of Data

By 'large' in this context I mean that the number of objects to be classified exceeds, say, two hundred, and may be several thousands. Demand for such classifications exists, and recent applications using the Rothamsted program have included classifications of (a) 400 strains of staphylococci and micrococci, (b) 630 quadrats of Nigerian vegetation, (c) 550 cases of hearing deficiency in Jamaica, (d) 350 post-mortem examinations of calves. An expanding file of 20,000 soil profiles awaits suitable methods of analysis.

The study of large sets of data often reveals the inadequacy of a subdivision into 'clusters' as a means of summarizing the data. Whereas a small subsample of the data may suggest the existence of clusters, the implication that these clusters represent meaningful aggregations of objects may turn out to be false when further objects are added to the set. If we take as an example the stars in the night sky, the brightest stars have been clustered since ancient times into constellations, which are, on the whole, visible aggregations separated by dark empty spaces. It is well known that on a very clear night it is difficult to pick out the well known constellations among the background of intermediate stars of lesser magnitude. As a practical system the sky has now been divided into arbitrary regions corresponding to the ancient constellations, but for positional identification the stellar co-ordinates are more useful.

Natural clusters, if they exist, are regions of more than average density in some spatial representation of the data. Without a reasonably large number of objects, meaningful statements about density cannot be made. However, data in which totally distinct clusters can readily be recognised are seldom submitted to the expensive process of computer analysis.

I am not going to discuss divisive (monothetic) methods, although these can usually handle quite large sets of data. Their committal to division on a single character at each stage often involves too much loss of information to be of great use. Neither am I going to discuss the important topics of the choice of characters and computation of similarity (or distance) coefficients.

The most informative techniques are those that show the structure of the set of objects. From a structural representation of the data a satisfactory set of clusters can often be made by visual inspection. The structural representation also allows automatic clustering procedures to be interpreted and their adequacy to be assessed. Once the principle has been accepted that there is no unique system of clustering suitable for all applications, it becomes clear that clustering is often more an administrative convenience than a scientific reality, except in situations where it can be shown that no intermediate objects are likely to exist between two separate clusters.

Methods used at Rothamsted

The classification programs used at Rothamsted are primarily aids to visualising the structure of multivariate data. Given a data matrix a similarity matrix can be computed using a similarity coefficient that can handle any combination of dichotomous, qualitative or quantitative characters. For small sets of data (up to 115 objects) Gower's method of Principal Co-ordinate Analysis (Gower, 1966) can be used. Principal Co-ordinate Analysis is a latent root and vector method that provides co-ordinates for each object as a point in n-dimensional space, such that the distance between any pair of points is equal to $\sqrt{2(1-s)}$, where s is the similarity coefficient between the two corresponding objects. By considering only the co-ordinates corresponding to the few largest latent roots, a useful visual representation of the set of objects can usually be obtained.

A disadvantage of Principal Co-ordinate Analysis is that when the representation in three or four dimensions is not adequate, there may be apparent clusters that would split if further dimensions were considered.

The second method of structure determination is the use of the Minimum Spanning Tree (Gower and Ross, 1968). The Minimum Spanning Tree of a set of points is the network of minimum total length such that every point is joined by some path to every other point, and no closed loops occur. Exactly n-1 links are required to connect n points. Several methods of computing the Minimum Spanning Tree are known, of which the algorithm of Prim (1957) is the most efficient. The method known as Wroclaw Taxonomy (Florek *et al*, 1951) also uses the minimum spanning tree. In practice we use the inverse relationship between distance and similarity, and operate directly on the similarity matrix, handling up to 600 objects.

From the minimum spanning tree, the single linkage cluster analysis of Sneath can be computed directly (Gower and Ross, 1968). By omitting from the tree all links with similarity less than a given value the remaining links form chained clusters, which are the required single linkage clusters. We have routines that provide a direct line printer representation of the dendrogram for single linkage cluster analysis. These routines will be published in the Algorithms Supplement of Applied Statistics. (Ross, 1969).

The disadvantage of the minimum spanning tree is that it provides no information about how the various branches of the tree should lie relative to each other. This can be overcome for small trees by drawing the tree on the vector diagram provided by

Principal Co-ordinate Analysis. For larger trees a table of the few nearest neighbours of each object is very useful in planning how the branches should be drawn.

The minimum spanning tree also shows where outlying points should be placed if it is not desired to treat all such points as separate groups.

Extending the methods to larger sets

The size limitations on the methods described may of course be modified when faster and larger machines can be used, but it is worth considering whether approximately equivalent results can be obtained with much less computation. Clearly a large similarity matrix contains much redundant information for Principal Co-ordinate Analysis if a low order representation is possible, and similarly the minimum spanning tree requires only a fraction of the links stored in the similarity matrix. One obvious redundancy is that if two points are identical their similarities with any third point will be equal, and if they are nearly identical the similarities will be nearly equal. However, it is difficult to be certain *á priori* which similarities are required and which are not.

The most promising line of attack is to select a base set or reference set of points to which all other points are then referred. Gower (1968) showed how an extra point can be fitted into a Principal Co-ordinate Analysis, and there is no reason why this should not be done for all points not included in the reference set. The resultant vectors may or may not correspond closely with the largest vectors of the full matrix had it been calculated. The conditions of success have not been fully investigated but it seems that, if (i) the reference set is representative of the full set in the sense that no large groups occur outside the region spanned by the reference set, and (ii) a good representation may be found in a few dimensions, then a recognisably similar diagram will appear, subject to rotation and relocation of the origin. The clusters formed in the vector diagram could then be re-investigated. The advantage of this method is that only the first few columns of the similarity matrix are used, and these may contain sufficient information for a satisfactory classification.

The minimum spanning tree can be computed for larger sets than at present if there is no need to store the similarity matrix. Because the data matrix usually takes up less space than the similarity matrix for large sets, and Prim's algorithm requires the elements of the matrix once only, less space (but not time), is needed for the computation.

Alternatively we have to select a reference set as before and attempt to find an approximation to the minimum spanning tree without working out too many unnecessary similarity coefficients. The method in outline is as follows:

(i) Select a reference set of r points.

(ii) Calculate the similarity of each point with each reference point and assign each point to one or more of r groups, *i.e.* to the groups founded by the reference point to which it has closest similarity, and to any other groups for which the similarity is within a small percentage, e, say, of this similarity.

(iii) Compute the minimum spanning tree for the members of each group, and for the members of the reference set.

(iv) Merge all the resulting trees (which have some points in common) to form the minimum spanning tree for the whole set. The resulting tree is minimal for the matrix defined by treating all uncomputed similarities as unknown.

The success of this method depends on the idea that the space can be divided into overlapping regions and all or most of the relevant links will lie totally within one of these regions. Thus two highly similar objects will probably be assigned to the same group and their link will be correctly included. The successful placing of outliers is less certain but it is also less important.

Note that the groups so defined are not clusters in any taxonomic sense but are purely a device to lessen computation.

The merging procedure will sort out any anomalies caused by the choice of reference points and the final tree will be treated exactly as if it had been computed in full.

Simple calculations suggest that, when the threshold is chosen so that about 30% of points are assigned to more than one group, optimal efficiency is obtained by choosing roughly $\sqrt{N}$ reference points, where N is the total number of points, when the efficiency will be about $4/\sqrt{N}$. Hence the time required depends on $N^{3/2}$ instead of N^2 as for the full analysis. Thus 10,000 objects could be classified using only 4% of the similarity matrix, equivalent to a full analysis of only 2,000 objects.

As an example, a known minimum spanning tree of 315 objects was compared with the trees obtained using the above method on (a) 21 systematically selected reference points, (b) 16 systematically selected reference points and (c) 16 reference points chosen from inspection of the data, with the following results:

	Full Analysis	(a)	(b)	(c)
No. of objects	315	315	315	315
No. of groups	1	21	16	16
No. of similarities used	49770	11792	11882	10512
Mean similarity % in M.S.Tree	57.3	53.7	54.2	55.0
% correct links	100	57	62	70
% second best links	0	14	15	13
% inferior links	0	29	23	17

As the mean similarity in the matrix is unusually small it is expected that results would be much better for matrices with greater mean similarity, but this has not yet been studied. However, when three-quarters of the information may be obtained from one-quarter of the work there may well be applications where the remaining information is not worth the expense of obtaining it.

References

Florek, K., Lukaszewicz, J., Perkal, J., Steinhaus, H., & Zubrycki, S. (1951). Taksonomia Wraclawska, Przeglad Antropologiezny, XVII, *193-207*. Poznan.

Gower, J.C.(1966). Some distance properties of latent root and vector methods used in multivariate analysis. *Biometrika, 53*, 325-338.

Gower, J.C. (1968). Adding a point to vector diagrams in multivariate analysis. *Biometrika* (In press).

Gower, J.C. & Ross, G.J.S. (1969). Minimum Spanning trees and single linkage cluster analysis. *Applied Statistics*. (In press).

Prim, R.C. (1957). Shortest connection matrix network and some generalisations. *Bell System Tech. J. 36*, 1389-1401.

Ross, G.J.S. (1969). Algorithms A513, A514 and A515. *Applied Statistics, Algorithms Supplement*.

Discussion

Q. (Parker-Rhodes) What criteria do you use for a set of data being unclassifiable? The examples of the constellation suggests that this can be highly controversial. Have you compared your results with data of non-statistical techniques?

A. You can consider data unclassifiable if you apply a single linkage cluster analysis and find no large divisions between groups but rather a single group which gradually accumulates more and more outliers. In such a case your classification into a set of small groups will not necessarily be very meaningful. However it may be very useful for administrative reasons to subdivide the minimum spanning tree with the realisation that you will have had to separate various pairs of objects which are in fact very close, because you wish to draw the line somewhere. You may want to make your groups of the same size (*e.g.* for batch processing) and as homogeneous as possible within the group, but you cannot expect all your groups to be equally homogeneous. In practice we have no definitive rules about what constitutes a good clustering. The minimum spanning tree often suggests a useful clustering by dividing up branches at their longest links and including outliers along with their nearest neighbours.

Q.(Paton) Do you have a program to find an adequate set of reference points? If not, what methods do you use for separating anomalous clusters up into their sub-clusters?

A. We do not have any standard procedures for this. Our experience so far suggests that as long as the reference points are not too close together (in which case you get considerable redundancy) or too far from the main body of data (in which case no points are sufficiently similar to them for them to be of much use), the method works reasonably well, and several different reference sets taken systematically or at random still give much the same answers because what you are trying to do is to find, if possible, the minimum spanning tree which approximates as closely as possible to that obtained by a full analysis. The clusters which are used for the purposes of the algorithm need not be useful clusters for other purposes.

Comment: (Hope) I suggest that whenever a particular level of a dendrogram is being printed out some measure of dispersion of each cluster should be introduced so that you can look at any sudden changes in that measure particularly when two clusters amalgamate. This became obvious when I applied taxonomic programs to well-known data by Fisher - his data on two sorts of irises which he published in 1936 to illustrate discriminant function analysis. And the program reconstituted the two sets of fifty irises perfectly, but the point at which you had fifty flowers in one cluster and fifty flowers in the other and these were the only two clusters, was signalled by the sudden increase in the dispersion of the conjunction set of all hundred flowers as compared to dispersions of the two individual subsets. This data is not perhaps the best data on which to illustrate this sort of analysis, because you have two very distinct clusters to start with, but I would put in a plea for someone to draw up some sets of validation data of different sorts, so that different taxonomic programs can be run on these data to compare the results.

Q.(Sneath) Do you have any experience of doing an R-type analysis on relatively few characters of numerous OTU's and automatic plotting of loadings of the OTU's on a few axes? This allows major grouping to be recognised by eye, but I think there is a danger that small but interesting clusters might not be recognised. Of course such an approach should be continued by doing thorough cluster analysis of each of the major groupings.

A. We have very little experience of using the method of adding a point in Principle Co-ordinate Analysis - we mean to study this. The idea is that having produced these diagrams you would then, as you suggest, select regions which suggest clusters and then re-analyse these smaller regions completely to get a much clearer idea of the detailed structure. Since you are looking only at the left hand strip of the S-matrix, you have no knowledge of any high similarities within the remaining uncalculated portion except for values inferred by some high measure of agreement between calculated rows of the matrix (that is, two close points will probably have equal links with any other member, although the converse does not always hold). Therefore the only way to find these close links, if they exist, is by more detailed analysis of the suggested groups. This is what we are trying to do in one complete process by the minimum spanning tree method. We assign each point to an appropriate region of the space, describe each region fully, and then link together the various regions.

Comment: (Jeffers) We have some experience of this method on sets of OTU's ranging from 1,500 to 7,000 in number, and with 8 - 30 characters. The results are extremely encouraging and, surprisingly, the small clusters are high-lighted, particularly if minimum spanning trees are computed for the subjectively chosen clusters.

AN ALGORITHM TO CONSTRUCT A PARTICULAR KIND OF HIERARCHY

M. Roux

Laboratoire de Biologie vegetale, Faculte des Sciences, Orsay

This talk is based upon an article of C.J. Jardine, N. Jardine and R. Sibson. The structure and construction of Taxonomie hierarchies in Mathematical Biosciences pp. 173-179 (1967) but I shall not summarize it here, I shall only take out of it what is useful for my purpose.

I must first speak about hierarchies and for that it is better to choose an example like the following one, in which are seen different levels, namely, a,b,c,d,e.

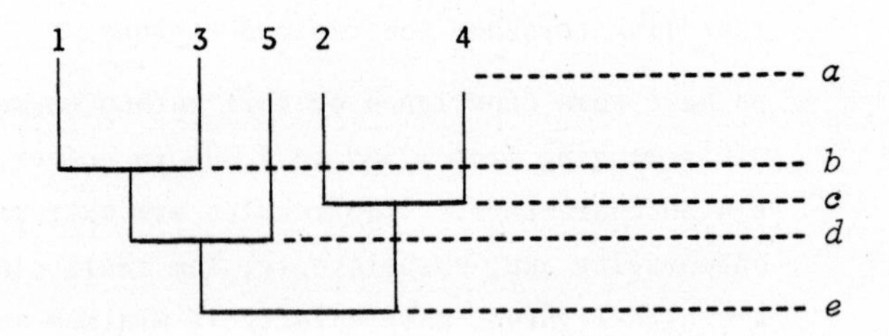

Usually these levels are increasing from e to $a = 1$, but I prefer to put $a = 0$ and the other numbers going increasingly to e, in such a way that I get a distance, in the mathematical sense, on the set of the five objects numbered here. In fact, the distance between two objects is obtained when looking up to the higher level you need to go down, to join the two objects under consideration. That is to say for example, the distance between 1 and 5 is d. Then I can construct a matrix of distances, if $a = 0$, $b = 1$. $c = 2$, $d = 3$, and $e = 4$.

	1	2	3	4	5
1	0	4	1	4	3
2		0	4	2	4
3			0	4	3
4				0	4
5					0

It is known that this is a distance and besides that it is an ultrametric one.

An ultrametric distance is a distance which satisfies the inequality (1):

$$\delta(i,j) \leq \max\ [\delta(i,k),\ \delta(j,k)] \text{ for all } i,j,k.$$

It is easy to see that this inequality implies that all the triangles have the two greater sides equal.

Conversely if we have a distance on a set of objects which satisfies the inequality (1) then it is possible to draw a taxonomic tree out of it.

We must now define an order on the set of all the metrics relative to a set J of objects: a metric d is greater than a metric d' if we have:

$$d(i,j) \geq d(i,j) \text{ for every pair } (i,j) \text{ of objects}$$

belonging to J. Suppose now we have a family $(d_s)_{s\varepsilon S}$ of distances (metrics) on a set J. We shall say that this family is bounded if for each pair (i,j), i and j being elements of J, the set $(d_s(i,j))_{s\varepsilon S}$ of real positive numbers is a bounded one. Then it can be shown that the application d^*, from J to JR^+ (the real positive numbers), defined by:

$$d^*(i,j) = \sup_{s\varepsilon S} [d_s(i,j)] \text{ is a metric on } J.$$

Furthermore, if the family $(d_s)_{s\varepsilon S}$ is a family of ultrametric then d^* is also an ultrametric.

We go back to our classifications: let J be a finite set of objects to be classified, and let us select a distance d on this set. We consider the family $(d_s)_{s\in S}$ of all the ultrametrics less than d. Obviously this family is bounded by d itself, so there exists a d^* greater than all the d_s but less or equal to d. The authors of the article cited, named it the subdominant ultrametric of d. In some sense d is the best ultrametric you can get from d.

Description of the algorithm

At the beginning of this algorithm we do all the $d^*(i,j)$ equal to the given $d(i,j)$ and after that, an iterative procedure takes place which consists of examining all the possible triangles and modifying them in such a way that the greatest side is reduced to be equal to the second side in size. Of course, if the triangle considered at one step has already the two greatest sides equal to each other there is no modification. And when all the triangles have been examined this way, they are examined again, until there can be an examination of all the groups of three points without further modification. This re-examination is necessary because a modification at one step may imply another modification in a triangle already examined.

In that way we obtain the subdominant. In fact, at each modifying step the d^* is always between the subdominant and the given metric. That is true at the beginning since then, the d^* is set equal to the d-distance. Now if at one step this is true then, at the following step this is true again: suppose we are examining the three points i,j,k having:

$$\delta(i,j) \le d^*(i,j)$$

$$\delta(i,k) \le d^*(i,k)$$

and $$\delta(j,k) \le d^*(j,k)$$

(I name δ the subdominant we are seeking for) and suppose that we have:

$$d^*(i,j) \geq d^*(j,k) \geq d^*(i,k)$$

(This does not remove any generality to the demonstration) then, after this step we will have $d^*(i,j) = d^*(j,k)$. But we had

$$\delta(i,j) \leq \max\ [\delta(i,k),\ \delta(j,k)] \leq d^*(j,k) = d^*(i,j)$$

Besides this procedure is finite because the set J to be classified being finite, the number of different measures $d^*(i,j)$ is finite and the number of possible modifications of one measure is also finite for, a modification can only reduce a measure to be equal to another one, and then the number of all the possible modifications is finite. And because of the test, the result is necessarily the subdominant.

Example

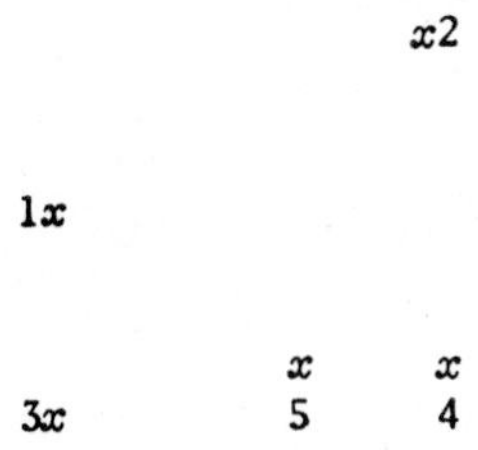

Fig. 1

	1	2	3	4	5
1	0	7.1	1	7.1	3.2
2		0	7.2	2	4.3
3			0	7	3
4				0	4
5					0

Fig. 2

	1	2	3	4	5
1	0	4	1	4	3
2		0	4	2	4
3			0	4	3
4				0	4
5					0

Fig.3

Fig.2 shows an approximation of the usual Euclidean distance between the points of fig.1. The first step of the algorithm consists of considering the triangle 1,2,3. Its sides are

(1,2) with length 7.1
(1,3) " " 1.
(2,3) " " 7.2

Then the modification is to reduce (2,3) to have 7.1 for length and so on. When arrived to the triangle 3,4,5 it is necessary to come back to the 1,2,3 and to do so until there is no more modification. Then the result is the Fig.3 which corresponds to the hierarchy tree selected as example at the beginning of this talk.

Conclusion

There are two advantages to this algorithm. The first is that it needs a small computer memory: only the matrix of distances to be transformed into an ultrametric, but the result is not clear enough because it is necessary to translate the obtained ultrametric into an hierarchic tree. The second advantage, which appears in all the trials we have done, is that it is very quick, generally it needs only four examinations of the whole set of the triangles. But there is a disadvantage: a very bad chaining effect.

ULTRAZ - A Fortran IV subroutine to get the subdominant ultrametric out of a given distance-matrix.

Author: M. Roux, Laboratoire de Biologie vegetale, Bat.490, Faculte des Sciences, 91-Orsay-France.

Storage: very small 50 Fortran IV statements. Other subroutine required: none.

Card decks, programme listing and any information may be obtained from the author.

Discussion

Q.(Wishart) In your example you used Euclidean distances. Have you also applied your algorithm to non-metric measures?

A. I have tried to use non-metric measures and the results are quite good but my experience is very limited.

Comment: (Lerman) I should like to note that I developed the same algorithm independently of Roux, reasoning only from the pre-order on the set of unordered pairs of objects, and I expressed the optimal property of this algorithm in terms of the proximity between two pre-orders. The algorithm, in practical terms, is analogous to Sneath's Single Linkage method but more systematic because the levels of grouping are not chosen *á priori*.

Q.(Parker-Rhodes) How does the computation time vary with the number of elements in the universe?

(Ross) The method described appears to be identical to Sneath's Single Linkage Method, except that the point at which two clusters merge is measured exactly rather than at uniform intervals. Since the method requires triplets of elements to be examined it must depend upon the cube of the number of elements. It would seem preferable to use methods which depend only on the square.

(Tomassone) One of the most important aspects underlying this algorithm is the fact that it is distribution free. This aspect is important because the distributions underlying the data are seldom known.

(Sneath) Listing the resemblance coefficients in rank order of magnitude is very convenient if one wishes to do Single Linkage on small samples by hand. The method used in the computer depends upon its speed, storage capacity, etc., and is a technical problem rather than one of theoretical interest in taxonomy.

Q.(Jeffers) Have you submitted any random data to the algorithm to see what sort of classification it produces?

A. No.

Q.(Jeffers) I suspect it would produce a classification even where none was present.

(Bisby) Any clustering programme will cluster random numbers. Some assumption about the value of the data is required before starting the analysis.

Q.(Jackson) Jardine has a function which measures the strain between his classification and his data and as far as I can see from your analysis you have no such measure of strain and your algorithm is in fact a sorting strategy whose mode is governed by your ultrametric relationship. I therefore want to ask why do you call it a classification strategy?

A. (After some discussion) Excuse me because I forgot to point out that there exists only one way to go from hierarchy to ultrametric and conversely only one way to go from ultrametric to hierarchy.

APPLICATIONS OF CROSS-ASSOCIATION TO AN EVOLUTIONARY STUDY OF CYTOCHROME *c*

M. J. Sackin

Microbial Systematics Research Unit, Leicester

Introduction

The purpose of this presentation is to show the potential of amino acid sequences of proteins for classification of organisms and for reconstruction of their evolutionary paths. At present there seems to be no known satisfactory way of picking out operationally homologous sets of characters from the sequences suitable for a proper numerical taxonomic study, though the cross-association method gives some help in this. The result is a more *ad hoc* method of working.

The work reported here is on the protein cytochrome *c*. Evolutionary studies on this protein are most illuminating, since the protein is found in most kinds of organism, and the cytochrome *c* sequence from over 20 animals, plants and fungi, and bacteria are known. This paper will show sequence similarities between certain bacterial cytochromes *c* and those of the more complex organisms.

The computer method used here derives from a joint study with Dr P.H.A. Sneath on methods of comparing protein sequences and other sequences of nonumeric data. The main program is an aid in reconstructing the evolution of proteins by the following generally accepted types of mutational event in the parent gene:

(i) substitution of a single nucleotide base where it shows itself as a single amino acid substitution in the protein. (The degeneracy of the amino acid code is such that certain base substitutions leave the corresponding protein unchanged);

(ii) deletion or insertion of three or a multiple of three bases, which lends to the deletion or insertion of one or more amino acids in the corresponding proetin. (Any other number of base deletions or insertions is called a "frame shift" mutation and is not revealed by our method).

(iii) duplication of a nucleotide chain within the gene, which leads to a repetition within the protein sequence;

(iv) duplication of the whole gene and transfer of one daughter gene to a different part of the genome, leading to two identical proteins which subsequently evolve independently.

The program gives numerical values to the similarities found and compares them with those that would be obtained if the two sequences were in random order.

Method

The method described here is called cross-association and is loosely analogous to cross-correlation. In cross correlation two sequences of numeric elements are compared by sliding one sequence against the other and finding the correlation coefficient between the overlapping portions at each step. In cross-association two sequences are also compared, but their elements are qualitative or non-numeric (*e.g.* amino acids). The coefficient of resemblance is not a correlation coefficient but an association coefficient in the sense in which this is used in Numerical Taxonomy (see Sokal and Sneath, 1963). It measures the number of elements which match at each overlap position (Sackin and Sneath, 1965). Cross-association has been incorporated into computer programs in ALGOL and FORTRAN IV (see Sackin, Sneath and Merriam, 1965, and Harbaugh and Merriam, 1968, where applications in geology for comparing rock sequences are described).

The cross-association program first reads in the two sequences to be examined. It then calculates the probability P that a randomly chosen pair of amino acids, one for each sequence, will match. It is noteworthy that among actual proteins P is remarkably constant. Over 31 pairwise comparisons of proteins the mean value of P is 0.0715 with standard deviation of only 0.0115. Some of the protein pairs were of the same family of proteins; others were of different families, between which high matching was neither expected nor observed. The value of P does not vary with the degree of similarity between the protein sequences compared.

The program next slides the two sequences past each other in steps of one amino acid. Suppose L, M are the lengths of the two sequences. Then there will be $L + M - 1$ distinct positions of overlap. For each position of overlap the program counts the number of amino acids that match and computes statistics indicating the degree of matching. In the following example high matching occurs in two successive overlap positions:

	Gly	*Ala*	*Glu*	*Asp*	*Lys*	*Phe*	*Gly*	*Leu*	*Asn*	*Ile*
	Gly	*Ala*	*Glu*	*Asp*	*Phe*	*Gly*	*Leu*	*Asn*	*Ile*	
Match:	Yes	Yes	Yes	Yes	No	No	No	No	No	

	Gly	*Ala*	*Glu*	*Asp*	*Lys*	*Phe*	*Gly*	*Leu*	*Asn*	*Ile*
		Gly	*Ala*	*Glu*	*Asp*	*Phe*	*Gly*	*Leu*	*Asn*	*Ile*
Match:		No	No	No	No	Yes	Yes	Yes	Yes	Yes

If these were actual proteins one might hypothesize that they had a common ancestor and have since evolved by the deletion or insertion of a single lysine residue in one of the two lines of descent. In support of this hypothesis one may cite the high matching in these two overlap positions. In general a deletion of j consecutive amino acids from one sequence (or insertion in the other) would produce high matching in two overlap positions j positions apart.

Table 1 shows the computer output for the two short sequences:

Ala Gly Lys Gly
Ala Thr Gly

SLIDING STEP = 1

PROB(MATCH) = .25000000

FORWARD MATCHES

MATCH POS.	NO. OF MATCHES	NO. OF COMPS	MATCHES/ COMPS	STD DEVS	CHI-SQ UNCORRECTED	CHI-SQ (YATES)
1	0	1	.0000	-1.0472	0.3333	0.3333
2	1	2	.5000	0.7405	0.6667	0.0000
3	1	3	.3333	0.3183	0.1111	0.1111
4	1	3	.3333	0.3183	0.1111	0.1111
5	0	2	.0000	-1.4810	0.6667	0.0000
6	0	1	.0000	-1.0472	0.3333	0.3333

SUM CHI-SQ (UNCORRECTED) = 2.2222, WHICH IS -1.2084 STD DEVS FROM THE MEAN (NORMAL APPROX).

Table 1. Sample cross-association computer output

The first column identifies the current overlap position. The second sequence may be regarded as sliding from left to right. Thus, the second line of the table (MATCH POS.2) refers to the sequences in the following alighment:

Ala Gly Lys Gly
Ala Thr Gly

The value of P is shown as PROB(MATCH) in the output. It is used to compare the matching at each overlap position with the matching expected by chance on the basis that the number of matches is a sample from a binomial distribution whose mean is P times the size of the overlap. For full details of the output see Sackin, Sneath and Merriam (1965).

For examining an overlap position with high matching a program has been written which will print out the two sequences under study in any "?" positions of overlap. It has been used here to look for grouping of matches along the overlapped chains.

Results

Similarities between cytochromes *c* of bacteria and of the more complex "eukaryote" organisms will be given, indicating a possible common origin for all these proteins. The high similarities among the known cytochromes *c* from eukaryote organisms themselves have been noted by many authors. See especially Margoliash and Smith (1965). In this class vertebrate, insect, plant and fungal sequences are known. Even between animal and plant or animal and fungal sequences about 60 per cent of just over 100 amino acids match when the two sequences are placed in a position of complete overlap. Such high resemblances strongly indicate a common origin for these proteins. Furthermore, the sequence differences among the proteins are roughly proportional to the differences among the organisms themselves as inferred from morphological and other characters. In the literature this type of cytochrome *c* has been referred to as the "mammalian type" cytochrome *c*. Clearly the use of this term is misleading, since it does not indicate the wide range of organisms whose cytochromes *c* belong to this class. The term "eukaryote cytochrome *c*" will therefore be used here to denote this class. If a eukaryote is found whose cytochrome *c* is not homologous with the other eukaryote cytochromes *c* then the term may need modification.

In the comparisons with bacterial sequences a hypothetical "average" eukaryote sequence has been used as the representative of its class. It is a reconstruction of the cytochrome *c* sequence of the most recent ancestor common to the vertebrates and the insects. It is the earliest sequence that can be reconstructed with worthwhile accuracy from the available data. It differs from each of the known animal sequences by about 15-20 amino acids and from the plant and fungal sequences by about 40 amino acids. However, the choice of eukaryote sequence makes very little difference to the results.

One of the first bacterial cytochromes *c* to be "sequenced", and still one of the few known bacterial sequences, is that from

Pseudomonas fluorescens. Neither visual examination nor cross-association has revealed similarities with eukaryote cytochrome *c*, though Cantor and Jukes (1966b) have proposed that both types of cytochrome *c* are in large part descended from an ancestral pentadecapeptide, with subsequent amino acid substitutions and a few deletions. Our studies on the whole complement those of Cantor and Jukes and provide links between eukaryote and *Pseudomonas* cytochromes *c* only via two other bacterial proteins, *Chromatium* cytochromoid and *Rhodospirillum rubrum* cytochrome *c*2. Though only partial sequences of these two proteins have been published (Dus and Kamen, 1963) it is apparent that both proteins exhibit strong sequence similarities to eukaryote cytochrome *c*, and the *Rhodespirillum* protein bears strong, varied and interesting relationships to the cytochrome *c* of *Pseudomonas*. The similarities, which are summarised in figure 1, will be discussed in turn.

General

As is well known, each of the proteins under consideration contains a peptide of the form *Cys-X-X-Cys-His* which binds a haem group. The *Chromatium* protein contains a further *Cys-His* peptide and a second haem group but its mode of attachment has not yet been fully determined.

Eukaryote-Chromatium

High matching occurs in two regions of the proteins. The first is as follows and corresponds to the lightly stippled areas in figure 1.

Residue No.	10					15				19
Eukaryote*	*Phe*	*Val*	*Gln*	*Arg* *(Lys)*	*Cys*	*Ala* *(Ser)*	*Gln*	*Cys*	*His*	*Thr*
	20					25				29
Chromatium	*Phe*	*Ala*	*Gly*	*Lys*	*Cys*	*Ser*	*Gln*	*Cys*	*His*	*Thr*
Min. base differences	0	1	2	1(0)	0	1(0)	0	0	0	0

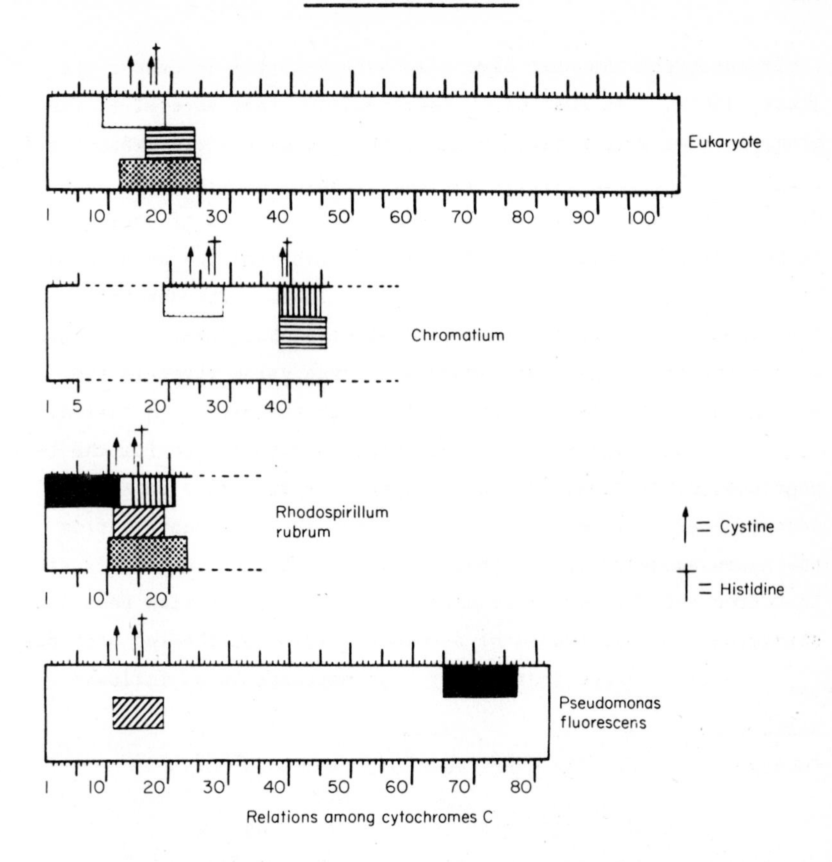

Fig. 1. The diagram shows maps of the four cytochromes c under consideration. Dotted lines indicate that the corresponding parts of the sequence have not been determined. Thus, the orderings of Chromatium residues 6-19 and Rhodospirillum residues 8 and 9 are unknown. In each case the residue numbered 1 is the NH_2 - terminal residue. Only NH_2-terminal parts of the Chromatium and Rhodospirillum sequences are known. Shaded areas represent six types of partially matching peptides as shown by the six types of shading.

(This match and the next have also been reported by Cantor and Jukes, 1966a. It must be further admitted that several of the other matches were noticed without the aid of the cross-association program. See also Dayhoff and Eck, 1968). The minimum base differences refer to the differences between pairs of base triplets in the parent nucleotide chains as inferred from the amino acid sequences from knowledge of the genetic code. The degeneracy of the code leads to uncertainty in some base assignments and hence in the numbers of base differences. Each value given is the minimum over all possible assignments for the pair in question.

Such high matching suggests a common origin for the two peptides, or at least for the rightmost heptapeptide in the scheme. It is interesting to note that the heptapeptide from the human sequence (*i.e.* residues 13-19) matches completely with that from the *Chromatium* sequence, but this observation does not strengthen the hypothesis of a common origin for the two peptides.

The second instance of high matching is as follows:

Residue No.	17			20				24
Eukaryote*	*Cys*	*His*	*Thr*	*Val*	*Glu*	*Lys* (*Ala*)	*Gly*	*Gly*
	39	40					45	46
Chromatium	*Cys*	*His*	*Thr*	*Phe*	*Asp*	*Glu*	*Gly*	*Ser*
Min. Base Differences	0	0	0	1	1	1(1)	0	1

This corresponds to the horizontally shaded areas in figure 1. The matching here is less striking than in the first scheme, although there is a minimum of only four base differences in the nucleic acids which code for the actapeptides.

* The residues shown in parentheses refer to sites in the ancestral eukaryote sequence whose residue assignment is tentative. They are assigned with lower probability than the residues shown immediately above them.

Eukaryote-Rhodospirillum. Matching between portions of these sequences is also very strong:

Residue No.	13		15					20					25
Eukaryote*	*Arg* (*Lys*)	*Cys*	*Ala* (*Ser*)	*Gln*	*Cys*	*His*	*Thr*	*Val*	*Glu*	*Lys* (*Ala*)	*Gly*	*Gly*	*Lys*
	11				15					20			23
Rhodospirillum	*Lys*	*Cys*	*Leu*	*Ala*	*Cys*	*His*	*Thr*	*Phe*	*Asp*	*Glu*	*Gly*	*Ala*	*Lys*
Min. Base differences	1(0)	0	2(1)	2	0	0	0	1	1	1(1)	0	1	0

These peptides correspond to the heavily stippled areas in figure 1. In the two peptides shown there is a minimum number of between seven and nine base differences in the nucleic acids which code for them.

Chromatium-Rhodospirillum. Perfect matching strongly indicates a common origin for two heptapeptides (vertically shaded in figure 1):

Residue No.	39	40					45
Chromatium	*Cys*	*His*	*Thr*	*Phe*	*Asp*	*Glu*	*Gly*
	15					20	21
Rhodospirillum	*Cys*	*His*	*Thr*	*Phe*	*Asp*	*Glu*	*Gly*

As in the other cases a common origin for these remarkably similar peptides seems more convincing an explanation than parallel or convergent evolution.

Chromatium-Pseudomonas. Comparisons of these two sequences have not revealed any particularly high matching.

Rhodospirillum-Pseudomonas. The following portions of the sequences match well (see diagonally shaded areas in figure 1):

Residue No.	12			15				19
Rhodospirillum	*Cys*	*Leu*	*Ala*	*Cys*	*His*	*Thr*	*Phe*	*Asp*
	12			15				19
Pseudomonas	*Cys*	*Val*	*Ala*	*Cys*	*His*	*Ala*	*Ile*	*Asp*
Min.base differences	0	1	0	0	0	1	1	0

* The residues shown in parentheses refer to sites in the ancestral eukaryote sequence whose residue assignment is tentative. They are assigned with lower probability than the residues shown immediately above them.

The identity of the residue numbers might lead one to expect high matching over the remainder of this overlap position. This is, however, not the case.

Consideration of other overlap positions has revealed further peptides which match well (black areas in figure 1):

Residue No.	1				5				10		12
Rhodospirillum	*Glu*	*Gly*	*Asp*	*Ala*	*Gly*	*Ala*	*Gln*	*(Thr,Leu)*	*Ser*	*Lys*	*Cys*
	66				70				75		77
Pseudomonas	*Val*	*Ser*	*Asp*	*Asp*	*Glu*	*Ala*	*Gln*	*Thr Leu*	*Ala*	*Lys*	*Try*
Min. base differences	1	1	0	1	1	0	0		1	0	1

(The ordering of Rhodospirillum residues 8 and 9 has not yet been determined, as far as is known to the author).
This comparison is particularly interesting since it is the only one which involves a part of the protein (in this case the *Pseudomonas* molecule) away from the haem-binding cystine and histidine residues.

Any other matching among the four proteins is less impressive than the cases given above. Thus in eukaryote cytochrome *c* the lightly stippled area (figure 1) overlaps with the horizontally shaded and heavily stippled areas, and one might expect the two latter areas to appear in *Chromatium* in the same positions relative to the lightly stippled area. Such matching, however, is not clearly found.

Discussion

A statistical assessment of the results is difficult in view of the complexity of the matching scheme and the attendant difficulties in setting up suitable null hypotheses. The statistical analysis of cross-association was utilised only in the early stages of the work, that is, for highlighting the overlap positions containing areas of high matching. Some matches would clearly be extremely unlikely events in random permutations of the sequences under study, for example the identical heptapeptide in the Chromatium and *Rhodospirillum* proteins.

The combined effect of all the matches is reinforcing and strengthens the claim that all four proteins derive from a common ancestor. It is, however, not clear by what mutational steps it evolved into the present-day proteins.

It is noteworthy that all but one of the six comparisons are of pairs of sequences in the neighbourhood of the haem-binding cystine and histidine residues. It is probable that for the molecule to function normally there are heavy constraints on the amino acids which can appear in these parts of the molecule. This may indicate that successful mutations in this part of the molecule have been comparatively rare and that the rate of evolution is lower than in other parts of the molecule. This further suggests a common origin of the proteins. On the other hand these same constraints may have encouraged rapid evolutionary convergence from different ancestral forms.

Regardless of the evolutionary interpretation of the results they give some of the first glimpses of a molecular homology between the proteins and hence the genes of eukaryote and prokaryote organisms. The "common origin" interpretation ties up well with current speculation of a common origin of prokaryote organisms and certain organelles which have been shown to be partially self-replicating systems with their own DNA complement (Gibor and Granick, 1964). Eukaryote cytochrome *c* occurs only in one of these organelles, the mitochondrion, but it is not known whether the protein is manufactured off the mitochondrial DNA. With the protein ferredoxin the situation is similar in that sequence similarities have been found between certain bacterial ferredoxins and the ferredoxin of a green plant (see Matsubara, Sasaki, and Chain, 1967). Plant ferredoxin is found only in the plastids but it is likewise not known whether it is coded by the plastid DNA.

There are two main problems in using protein sequences for performing a numerical taxonomy on organisms. The first, which in this paper has not been fully solved, is finding a general definition of an operational homology. The second is the existence of several molecular species of the same protein unit (*e.g.* haemoglobin β and γ) in the same organism. These units may have evolved from an ancestral form which existed prior to the time of the organism ancestral to those being classified. Their use as classification data may thus contribute towards a spuriously low similarity between the organisms (see Dayhoff and Eck, 1968, for a further discussion of this point).

However, protein sequences are of potential taxonomic use, especially for very distantly related organisms where an operational homology in morphological characters is hard to define. They should also be of use in classifying viruses whose morphologies are defined by very few parameters, and where there is also an acute problem of defining an operational homology for the characters.

Note added at proof stage

The complete amino acid sequence of *Rhodospirillum rubrum* cytochrome c_2 has now been published (Dus, K., Sletten, K., and Kamen, M.D., 1968. *J. Biol. Chem.*, *243*, 5507). Examination of this sequence strengthens the validity of the findings given in this paper, so confirming the predictive power of cross-association. However, in view of revisions to the sequence as previously published (Dus and Kamen, 1963) the matching of *Rhodospirillum* residues 1-12 with *Pseudomonas* residues 66-77 is no longer high.

Acknowledgements

This paper derives from work towards an M.Sc. degree at Leicester University which the author obtained while a staff member of the Medical Research Council Microbial Systematics Research Unit under the supervision of Dr P.H.A. Sneath who has given valuable assistance at all stages of the work.

References

Ambler, R.P. (1963). The amino acid sequence of Pseudomonas Cytochrome *c*-551. *Biochem. J., 89* : 349.

Cantor, C.R. and Jukes, T.H. (1966a). Repetitions in the polypeptide sequence of cytochromes. *Biochem. Biophys. Res. Commun., 23* : 319.

Dayhoff, M.O., and Eck, R.M. (1968). Atlas of Protein Sequence and Structure, 1967-68. National Biomedical Research Foundation, Silver Spring, Maryland.

Dus, K., and Kamen, M.D., (1963). Comparative structural studies on some bacterial heme proteins. *Biochem. Z., 338* : 364.

Dus, K., Sletten, K., and Kamen, M.D. (1968). Cytochrome c_2 of *Rhodospirillum rubrum.* II. Complete Amino Acid Sequence and Phylogenetic Relationships. *J. Biol. Chem., 243* : 5507.

Gibor, A., and Granick, S., (1964). Plastids and mitochondria: inheritable systems. *Science, 145* : 890.

Harbaugh, J.E., and Merriam, D.F. (1968). Computer Applications in Stratigraphic Analysis. John Wiley and Sons, Inc., New York.

Margoliash, E., and Smith, E.L. (1965). Structural and functional aspects of cytochrome *c* in relation to evolution. In Evolving Genes and Proteins. (V. Bryson and H.J. Vogel, eds.), Academic Press, New York, p.221).

Matsubara, H., Sasaki, R.M., and Chain, R.K. (1967). The amino acid sequence of spinach ferredoxin. *Proc. Nat. Acad. Sci. U.S., 57* : 439.

Sackin, M.J., and Sneath, P.H.A. (1965). Amino acid sequences in proteins: a computer study. *Biochem. J., 96* : 70P.

Sackin, M.J., Sneath, P.H.A., and Merriam, D.F., (1965). ALGOL program for cross-association of non-numeric sequences using a medium size computer. *Kansas Geol. Survey Special Distribution Publ. no. 23.*

Sokal, R.R., and Sneath, P.H.A. (1963). Principles of Numerical Taxonomy, W.H. Freeman and Company, San Francisco - London.

Discussion

Q.(Crawford) If cytochrome *c* from wheat and man shows an amino-acid similarity of 60% - is this not related to parallel evolution which in itself would be necessary to preserve the functional structure of vitamin C?

A. It is possible but it really seems to be too much of a coincidence.

Q.(Greenwood) The whole point about Natural Selection is surely that it is a systematic mechanism for producing things that would be exceedingly unlikely to occur by chance. Furthermore, if an organism requires a molecule it will evolve it. Let us take the case of haemoglobin which occurs in vertebrates, in some anthropods, in some annelids, in a few molluscs, and even in a few bacteria. Now, on the basis of all the other features of the organisms possessing haemoglobin it is most improbable that they have a common ancestry distinct from organisms without haemoglobin. We must realise that it is organisms which evolve and not molecules.

A. My point was that it is the genetic material which evolves, not the molecules in isolation, or do you disagree with that?

Comment (Greenwood) It evolves because the organism evolves, not in isolation.

A. I should have thought they were one and the same thing.

Comment (Greenwood) That is the mistake which the molecular biologists make.

Comment (Sneath) The difficulty of supporting convergence as the main explanation for the striking resemblances between these proteins is even greater than may appear. If the similar portions are of similar sequence simply because of functional requirements, then this implies that other, highly variable, areas have very loose functional requirements. This, of course, may be true, but we know very little indeed about functional requirements, and what little is known suggests that functional requirements of protein structure are scattered along much of the length of the sequences. Also, from our knowledge of haemoglobins it appears that they have arisen from the muscle protein, myoglobin. It would be quite plausible therefore if annelid haemoglobins have arisen from annelid muscle proteins, and perhaps all the haemoglobins could have evolved ultimately from a muscle-protein of primitive invertebrates.

Q. (Crawford) Cannot one have some test in molecular biology to distinguish convergence from other things, as we often have for morphological characters.

A. We have practically nothing to go on here.

Q. (Bisby) To what extent do you know whether the regions of cytochrome that you are dealing with are important in the cytochrome's function? This will determine whether or not the occurrence of matching sequences is a necessity for the protein to be extracted as cytochrome, or whether it is in fact an unlikely occurrence indicating possibly common descent.

A. Nobody knows really. There is a case for both points of view.

Q. (Craig) If you have two sequences say *ABCDEFG* and *ABCDXEFG* is the degree of similarity (or overlap) *ABCD* only or *ABCDEFG*? It seems to me that the overlap is greater than shown in your slide. In the two sequences effectively there is similarity of 7 characters.

A. I am sorry if I didn't make this clear. It is the strength of true matching in the two separate overlaps which determines how strongly you can call it a deletion or insertion - whatever I mean by that.

In the overlap position:

ABCDEFG
ABCDXEFG

the *EFG*'s match. In the next overlap position:

ABCDEFG
ABCDXEFG

the *ABCD*'s match.

EVALUATION OF CLUSTERING METHODS

P. H. A. Sneath

M.R.C., Microbial Systematics Research Unit,
University of Leicester

As new methods of cluster analysis are devised it becomes necessary to consider in some detail their aims. The general intent of cluster analysis is to reduce detailed information about the many to generalizations about the few, and it is thus similar in this respect to factor analysis although the method is very different. The problems of evaluating cluster methods are quite severe. No general symbolism is yet available, and this is much needed. The form of a classification depends on its purpose, and the usual cluster methods are directed toward producing general purpose classifications, and are thus polythetic in concept, although some of them, such as Association Analysis of Williams and Lambert (1959), are technically monothetic. What is now needed is closer attention to the exact aims of clustering methods.

In the following discussion an agreed similarity matrix is assumed. Most of the points will be illustrated by Euclidean models, though some methods do not employ Euclidean metrics. Even with the latter, however, some aspects can be illustrated by analogy in Euclidean spaces.

External criteria

External criteria of a classification usually depend both on the resemblance coefficients and the clustering method, and it is seldom easy to separate the two. They may depend on testing the agreement with single factors or attributes (*e.g.* soil type in ecology, or pathological process in clinical medicine), or more rarely by comparison with an agreed system of classes or taxa. The most detailed consideration of this has been that of MacNaughton-Smith (1965), but relatively simple statistical techniques can be quite effective, such as chi-square tests between the membership lists of two sets of classes. One serious weakness is that the external criteria may themselves be largely matters of opinion. Another approach is to appeal to further evidence, but there may also be difficulties in this (Sneath, 1967). Either new attributes or new O.T.U's may be examined, and the concordance between the classifications can then be studied. This becomes, in effect, a form of internal testing in most cases (see below), because the most satisfactory studies consist of splitting the available knowledge into subsets, and comparing the results based on these subsets.

Internal criteria

The best-known method of testing a cluster method against the similarity matrix from which it was derived, is the cophenetic correlation technique of Sokal and Rohlf (Sokal and Rohlf, 1962). By this each entry in the similarity matrix is paired with a value obtained from the dendrogram by finding the similarity level that links the pair of O.T.U's in question. The cross-product correlation coefficient is then computed between all these pairs. This is the cophenetic correlation coefficient r_c. It has several useful properties; it is unchanged by the addition of a constant to either matrix, or by multiplication of a matrix by a constant factor. It is thus unaffected by an even stretching of the scale of the dendrogram, or by a shift in the origin of this scale. Some generalizations about the commoner cluster methods may be of interest (these methods are described in detail in Sokal and Sneath, 1963).

The highest cophenetic correlations are usually obtained with the Average Linkage methods. Of these, the Unweighted Pair Group Method generally gives the highest values of r_c, followed by the Unweighted Variable Group Method. The Weighted Group methods come next. As one would expect, the constraints imposed by Variable Group methods, as opposed to Pair Group methods, cause the former to give rather lower r_c values, but the difference is usually quite small. The weighting of stems in Weighted Group methods may also be viewed as constraints which make them less well-fitting than the Unweighted Group methods. With unweighted linkage the similarity level on the dendrogram between a point and a nearby cluster corresponds to a shorter distance (on the average) than the actual distances between the point and the members of the cluster. With weighted linkage the similarity level tends to represent a still shorter distance, which is thus in even less good accord with the actual distance.

After Average Linkage methods the Complete Linkage analysis gives the next highest values of r_c, followed by Single Linkage. We do not have much experience with other cluster analyses, but one gets the impression that the monothetic Association Analysis performs about as well as Single Linkage in this respect, while Information Analysis (Williams and Lambert, 1966) is rather better than Single Linkage (Watson *et al*, 1966).

It should be noted that the cophenetic correlation is not necessarily the ideal criterion for a clustering method. Thus it seems likely (in theory at least) that a set of O.T.U's forming an elongated cluster might be better represented by Single Linkage than by Average Linkage, despite a higher value of r_c for the latter.

Another criterion for the way a dendrogram represents the similarity matrix has been proposed by Hartigan (1967). This is the sum of squared differences between the matched entries. Although it is not invariant to scaling or translation it can be partitioned to show the contribution to the misfit given by selected O.T.U's and (with suitable modification) by selected characters. Criteria based only upon the branching pattern, without reference to the distances between branches, have not been extensively studied, though the work of Camin and Sokal (1965) is a step in this direction; their value, however, may not be very great. What is much needed is a development of analysis of variance and convariance that would allow different components of interest to be estimated.

General concepts of clusters

In a broad sense clusters are thought of as collections of points which are relatively close, but which are separated by empty regions of space from other clusters. There are many ways in which clusters can be defined, and several of the general concepts are briefly discussed below. Some of these concepts overlap considerably, but a review may serve to point out various resemblances and differences. Useful sources of references are Sebestyen (1962) and Ball (1965).

We may first distinguish overlapping clusters from those which are non-overlapping and usually hierarchic. While overlapping clusters are useful in some fields, hierarchic clusters are generally the most useful. The convenience of hierarchies is widely recognised, and so there may be some danger that non-overlapping clusters may be imposed upon data when they are unsuitable for revealing the structure that is present. We have more experience with hierarchic methods than with overlapping ones, so I shall in the main discuss the former.

An obvious distinction is between the round, compact cluster and the long straggly cluster. Cattell and Coulter (1966) have formalized this distinction and proposed different terms for them. They call compact clusters "stats" and straggly ones "aits". It is often thought that clusters usually are roughly spherical (or should be). This, however, is often not so, and it can be argued that the widespread occurrence of character correlations (which are not easy to remove, even if they are suspected) will mean that elongated clusters will be found with most sorts of data. Imposing spherical clusters on straggly data is an unsuitable procedure, and techniques for finding both kinds are needed. Complete Linkage and Average Linkage analyses will demonstrate mainly spherical clusters, while straggly ones can be found by Single Linkage analysis. Sometimes a compromise is required, and for this some modification of Single Linkage is often employed, with facilities to prevent chaining from becoming too extensive (*e.g.* Willmott and Grimshaw, 1969; Wishart, 1969;

Carmichael *et al*, 1968). Intermediate forms of linkage have also been suggested (*e.g.* Sneath, 1966a). Most of these techniques are intended to reduce background "noise", or to prevent the effects of fortuitously high resemblances between certain pairs of O.T.U's.

Another set of general concepts is the way in which clusters are conceptualized. We may distinguish three main kinds, based on concepts of mass, density and network.

The idea of O.T.U's as points of unit mass in character space is a ready one. Perhaps the clearest form of this is where the original points are replaced successively by hypothetical new points representing the centroids of clusters and having a mass proportional to the number of O.T.U's in the cluster. Such models can be handled by imagining an attracting force acting between the points. These models can be extended to cover clusters of unusual shape; for example, a method for finding linear patterns using a gravitational model has been described (Sneath, 1966b).

Clusters can also be thought of as volumes of high density in the character space, and some computer programs employ this idea (*e.g.* Wishart, 1969). The concept of density may be rather ill-defined, for it may also imply the idea of envelopes of constant density around the cluster. Many density models have the disadvantage that some additional parameter must be given by the user, to determine how the density shall be calculated. The mass of the points is, as it were, diffused into the surrounding space. This is rather like throwing out of focus a photograph of the points; this can be done to any desired degree until the clusters become so diffuse as to be valueless, and the optimal diffusion may be very uncertain. Also, it may not be feasible to define densities if the points are very few.

The idea of a cluster being defined by an envelope of some given density is a natural development. It is implicit in many multivariate techniques which aim to define envelopes of one or two standard deviations, for example. These are not difficult to manage when the points are few, because variances can still be usefully calculated except for single points; the diffusion of mass is based on the assumption of a normal distribution. The concept of envelopes has its converse in the idea of planes of separation, *i.e.* of low density. Envelope models are not well-suited to overlapping clusters.

A different concept is that of a network of links between the points of a cluster. This approach is well suited to graph theory, and in thinking of this we should perhaps concentrate our attention on the links rather than on the points. The minimal spanning tree is a good example of such a network, and single link methods may be usefully considered as network models.

There are a number of special points that merit attention. One of these is the effect of a clustering method on aberrant O.T.U's that are scattered singly around the periphery of the space. There is a strong psychological urge to force these points into the clusters nearest to them, in an effort to reduce the number of classes to be recognized, and some clustering methods have this effect. Various forms of information statistic (and allied methods) have this tendency, and they are also susceptible to the exact composition of the set of O.T.U's employed (Watson *et al*, 1966; Hall, 1967). The information on single isolated O.T.U's tends to be "lost" among that on more numerous O.T.U's. It may be noted that the Weighted Average Linkage methods have, in comparison to the Unweighted forms, the opposite effect, since they tend to give relatively more importance to isolated O.T.U's during clustering. The weighted method can perhaps be regarded as a device that has the effect of equalizing the size of the samples of O.T.U's of each of the smallest clusters. To give a biological example, if one had three species, *A*, *B* and *C*, with ten O.T.U's from *A* and *B* but only two from *C*, the weighted methods would tend to treat them as if there were about seven O.T.U's from each species. The biologist would often prefer this, which provides an additional reason in favour of the weighted methods despite the rather lower cophenetic correlations given by them.

The discussion above has been mainly on agglomerative clustering methods. Divisive methods tend to behave erratically, (although association analysis is quite reliable, despite being monothetic). They seem to be rather prone to creating heterogeneous groups of aberrant O.T.U's, while the minimum variance methods can apparently give quite unacceptable results; this has been found so for Ward's method by Forgey (1964), and H.R. Sanders tells me (personal communication) that the cluster method of Edwards and Cavalli-Sforza (1965) shows the same drawback (see also Wishart, 1969).

In comparing monothetic methods with polythetic ones an additional problem arises. Monothetic methods may give some bad "misplacements" of O.T.U's because these O.T.U's happen not to possess the property used for the higher monothetic divisions. We have, therefore, to specify in some way what degree of major misplacement we can tolerate. A classification which grouped even a single butterfly with the bees would meet severe criticism, when one showed minor misplacements within either taxon might be acceptable. A system of weighting the misplacements should be developed.

Tests for the presence of clusters

It must be admitted that one of the biggest deficiences of cluster analysis is the lack of rigorous tests for the presence of clusters and for testing the significance of those that are found. Although some criteria have been proposed (*e.g.* Goodall, 1966a, 1966b) there has not been any thorough study of this, especially in view of the many kinds of cluster we may recognize, and which might be present in our material. It would be extremely useful, too, to have a test to tell us when it was likely to be profitable to make a cluster analysis, or whether some other approach should be tried (such as factor analysis).

The general idea of the "clusteriness" of a set of points is related to the concept of entropy. The analogy is not too close, because entropy measures disorder, and both a regularly spaced distribution and a clustered distribution are ordered and thus have low entropy. Also, it is not widely recognized that the degree of clustering depends on the scale of observation of the space. Ecologists have been aware of this for many years, and some examples are given by Grieg-Smith (1964). One can, for instance, have a distribution consisting of regularly-spaced clusters, with a random distribution within each cluster. At different levels of testing this might appear random, clustered, or regular. Some methods based on the distribution of distances from an O.T.U. to its k-th neighbour have been proposed in ecology (*e.g.* Thomson, 1956) but this has not yet been developed in numerical taxonomy. Ideally such a technique would give a curve of the degree of clustering at different levels of testing, so that problems such as the one above could be readily solved.

Perhaps the most difficult problem is to set up satisfactory null hypotheses. This has been discussed elsewhere (Sneath, 1967), but a few points may be repeated. A random distribution of O.T.U's in phenetic hyperspace seems to be the most generally useful null hypothesis, and this is different from the common assumption of multivariate statistics that a multivariate normal distribution is appropriate. It leads to testing against the hypothesis that no clusters are present, instead of one cluster. Random distribution of character state values can give randomly placed O.T.U's, but it should be noted that this is not always so; for example the random choice of 1 and 0 values for binary characters will lead to something more like a multivariate normal cluster if the frequency of 1 states is the same for all characters. A random distribution of similarity values is not a realistic model, since it could scarcely occur in practice; it would mean that two O.T.U's that were identical would have quite different relationships to other O.T.U's. It has been suggested (Sneath, 1967) that use could be made of the dendrogram using Kolmogorov-Smirnov statistics, and this has the advantage that any null-hypothesis about the pattern can be readily utilized.

Computer feasibility

The practical point of whether a cluster method is feasible depends very much on the kinds of computer available today and in the future. It should not be overlooked, however, that any method which requires the testing of all possible combinations of clusterings will usually be impracticable for any but very small numbers of O.T.U's. Most commonly used methods require the calculation of the entire similarity matrix, but relatively little additional computation during clustering, and the time required is roughly proportional to the square of the number of O.T.U's. Active interest is now being shown in finding even quicker methods for very large numbers of O.T.U's, which avoid calculation of the entire similarity matrix. The development of very high speed computers in the future may entirely alter the outlook, and allow practical use to be made of some methods that are desirable but as yet scarcely practicable.

References

Ball, G.H. (1965), Data analysis in the social sciences: what about the details? *Proc. Fall Joint Computer Conf., 1965*, 533-559.

Camin, J.H. & Sokal, R.R. (1965), A method for deducing branched sequences in phylogeny. *Evolution 19*: 311-326.

Carmichael, J.W., George, J.A. & Julius, R.S. (1968), Finding natural clusters. *Systematic Zool., 17* : 144-150.

Cattell, R.B., and Coulter, M.A. (1966), Principles of behavioural taxonomy and the mathematical basis of the taxonome computer program. *Brit. J. Math. Statist. Psychol., 19* : 237-269.

Edwards, A.W.F., and Cavalli-Sforza, L.L. (1965), A method for cluster analysis. *Biometrics, 21* : 362-375.

Forgey, E.W. (1964), Evaluation of several methods for detecting sample mixtures from different N-dimensional populations. *Amer. Psychol. Assoc. Meetings, Los Angeles.*

Goodall, D.W. (1966a), Hypothesis-testing in classification. *Nature 211* : 329-330.

Goodall, D.W. (1966b), Numerical taxonomy of bacteria - some published data re-examined. *J. Gen. Microbiol.* *42* : 25-37.

Grieg-Smith, P. (1964), Quantitative Plant Ecology 2nd ed. Butterworth, London, 256 pp.

Hall, A.V. (1967), Studies in recently developed group-forming procedures in taxonomy and ecology. *J. South Afr. Bot.* *33* : 185-196.

Hartigan, J.A. (1967), Representation of similarity matrices by trees. *J. Amer. Statist. Ass.*, *62*: 1140-1158.

MacNaughton-Smith, P. (1965), Some statistical and other numerical techniques for classifying individuals. Home Office Research Unit Report. H.M.S.O., London, 33 pp.

Sebestyen, G.S. (1962), Decision-making processes in pattern recognition. Macmillan, New York, 162 pp.

Sneath, P.H.A. (1966a), A comparison of different clustering methods as applied to randomly-spaced points. *Classification Soc. Bull.*, *1* (No.2) : 2-18.

Sneath, P.H.A. (1966b), A method for curve-seeking from scattered points. *Computer J.*, *8* : 383-391.

Sneath, P.H.A., (1967), Some statistical problems in numerical taxonomy. *Statistician*, *17* : 1-12.

Sokal, R.R. and Rohlf, F.J. (1962), The comparison of dendrograms by objective methods. *Taxon*, *11* : 33-40.

Sokal, R.R. and Sneath, P.H.A. (1963), Principles of numerical taxonomy. W.H. Freeman, San Francisco, 359 pp.

Thompson, H.R. (1956), Distribution of distance to n-th neighbour in a population of randomly distributed individuals. *Ecology*, *37* : 391-394.

Watson, L., Williams, W.T. and Lance, G.N. (1966), Angiosperm taxonomy: a comparative study of some novel numerical techniques. *J. Linn. Soc. (Bot.)*, *59* : 491-501.

Williams, W.T. and Lambert, J.M. (1959), Multivariate methods in plant ecology. I. Association-analysis in plant communities. *J. Ecol.* *47* : 83-101.

Williams, W.T., and Lambert, J.M. (1966), Multivariate methods in plant ecology. V. Similarity analyses and information analysis. *J. Ecol.*, *54* : 427-445.

Willmott, A.J. and Grimshaw, P.N. (1969), Cluster analysis in social geography. This colloquium, pp.272-281.

Wishart, D. (1969), Mode analysis: a generalization of Nearest Neighbour which reduces chaining effects. This colloquium, pp.282-311.

Discussion

Comment: (Orloci) What did you mean by saying that a measure to be meaningful as a similarity or dissimilarity measure has to be locally euclidean?

A. I'm not sure that this is true in every case. But the difficulty is that if you have a sufficiently close triplet of objects and you are able to say that *A* and *B* are close and *C* is a long way away (which is effectively saying that it is at least metric - perhaps non-euclidean) what you are really saying is that you cannot measure the resemblance between *A* and *B* and *C* and say that one is greater than the other; this means that you have really destroyed the basis of your resemblance because you now can't put a rank order on the resemblances between *A*, *B* and *A*, *C*.

Comment: (Orloci) When you use any similarity measure, you may say that the elements of that measure are values of a metric, for example, correlation coefficients can be created from distances. But even in this case, it doesn't need to be; for example with information theory measures I do not think that you need to think of a metric at all, yet these measures are perfectly suitable to describe the relatedness of the individual objects.

A. I am not arguing here that you should use a metric - I was simply saying that I would use it for illustration. I don't know whether any of the resemblance measures which you use are not approximately euclidean when you come to very small distances, but I suppose that if you have an identity or almost an identity, whatever technique you use, you will bring the objects very close together. In this sense, the measures behave as though they were metric.

Q.(Cole) Could not a single linkage network density be defined as a curvilinear density? This could be extended to multi-linkage methods to a smallest path density, but this would involve considerable computational difficulties.

A. It should be worthwhile considering. Dr Jeffers has used the broken-stick statistic on prim networks as a kind of index of overall clusteriness. It does seem to work very well.

Comment: (Jeffers) This is a modification of the minimum-spanning tree. One compares the lengths into which the tree is broken with the lengths one would get by dropping a glass rod on the floor and breaking it into a similar number of pieces. The advantage of this is that it shows up the sizes at which the discontinuities are occurring in the multivariate space.

Comment: (Lerman) I think that it is necessary to find a relevant statistic which measures the ability of the set to be classified. I have proposed one criterion based on the pre-order on the set of the pairs for which one pair precedes another if the similarity of the first pair is less than that of the second. This criterion measures the proximity between the structure of the given pre-order and the structure of an ultrametric pre-order, and a test of significance is proposed with this statistic under a probabilistic hypothesis. Reference: I.C. Lerman, 'Analyse du Probleme de la recherche d'une hierarchie de classifications', Center de Calcul - H.S.H., Paris.

Q.(Crawford) Is the last diagram you showed the type of diagram used in Grieg-Smith's work in pattern analysis?

A. Yes. As one of my ecological friends pointed out, trees in the forest are relatively irregular where you have all kinds of scale factors of clusteriness. There is a technique known as nearest-neighbour or kth-neighbour search which can be used for testing for regular clusters in such situations, but I'm not sure if it has been suitably generalised for numerical taxonomy.

Q.(Jackson) Would you comment about the relationship between the stability of a classification with respect to small changes in the data, and the use of a metric.

A. I have not experimented with this except with bacteriological data where one or two extra bacteria are inserted to see what happens. All I have is an intuitive feeling that these groups were fairly robust under this treatment. Rohlf is publishing a paper showing the effects of mis-codings, and his conclusion is that they are still pretty stable -perhaps rather more stable than you would think.

Comment: (Barrs) Due to various editing reasons in a few of my O.T.U's over half of the data is missing. Not only do these O.T.U's get placed correctly in a multistate method with facilities for discounting non-recorded data, but when coded in binary form for programmes without these facilities, they still get placed in approximately the correct position.

Q.(Ord-Smith) Dr Sneath has said that often the human eye can do better than computer methods when detecting shapes. It occurs to me that here is an opportunity to tie the computer analysis in with on-line visual displays so that you could rotate dimensions, or look at planes through many dimensions. Has anything been done in this direction?

Comment: (Kelly) In answer to this, a paper was presented at IFIP (Edinburgh, 1968) by David Hall of Stanford, which described a system that does precisely this. (Reference: D.J. Hall, G.H. Ball, D.E. Wolf and J.W. Eusebio, Promenade - an interactive graphics pattern - recognition system, *IFIP Proceedings, 1968*).

Comment: (Parker-Rhodes) On the whole, you've made rather too much of this metric business. In particular the curved sausage is an artifact of the metric space, and it is just as easy to see in a non-metric space as in any other. I suspect that clumping methods, which are strictly non-metric, would be able to cope with this rather more quickly than anything in the metric space.

A. I hope you will experiment with this because we need all the help we can in coping with awkward cases. I don't have a vested interest in metrics as such, it is just that they are very easy to discuss.

Comment: (Willmott) Although much data is a heterogeneous mixture of parametric and non-parametric forms, and consequently it is then convenient to begin a discussion of this kind from the similarity matrix, consideration of binary data allows one to consider this problem from the original data. (Mr Willmott now gave an example of a Z-diagram on the blackboard).

CLUSTER ANALYSIS IN SOCIAL GEOGRAPHY

A. J. Willmott* and P. N. Grimshaw**

*Department of Computation, University of York

**Department of Geography, St. Helens College of Technology

Synopsis

The paper describes a single link taxonomic analysis of a data set consisting of over five hundred householders of a North Staffordshire town, each householder described in terms of his "mobility characteristics". The work sought to ascertain whether any underlying patterns of movement could be associated with the groups of householders drawn together by a taxonomy. A number of distinct groups emerged in the analysis, the associated movement patterns of which suggested the mobility, or lack of it, was strongly correlated with age, wealth, possession of a motor car and whether the householder was a "local" of one or more decades standing. "Basically it is the great social problem of the form and organisation of towns and cities that requires attention, and there is much to be done". (C. Buchanan 1963)*

Introduction

The Biddulph valley of Staffordshire, on the borders of the south-west Pennines and the Cheshire plain, is an area where physical and cultural landscapes are closely and clearly related. A triangular area is formed by a pitching syncline which drops south from The Cloud (1075') which forms the apex. The arms of the triangle consist of the Millstone Grit faulted, hogs-back of Congleton Edge in the west and the more gentle slopes of Biddulph Moor in the east. Near the centroid of the triangle lies the town of Biddulph about which this work is concerned. (Fig.1).

* C. Buchanan, Traffic in Towns (1963), 200

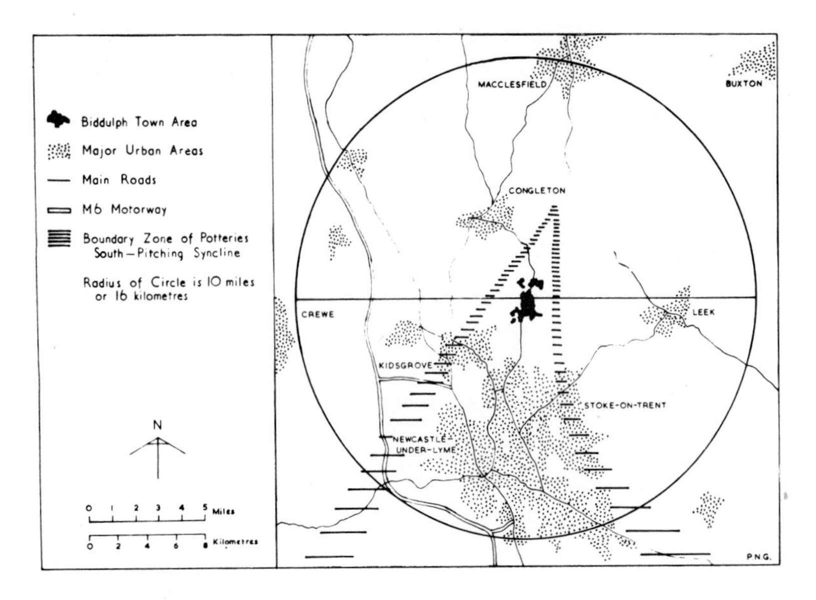

Fig. 1.

Communications with the Potteries in the south have developed easily whereas travel to Crewe and Kidsgrove in the west and to Leek in the east, for example, has been hampered by the physical geography of the arms of the triangle.

The establishment of basic local industries assisted Biddulph to maintain detachment from the nearest Pottery town. Indeed, engineering, coal and textiles developed in Biddulph account for nearly two thirds of the employed population of the Urban District. Since 1945 there has been a 50 per cent increase in population in Biddulph because of three factors, (i) a large housing scheme developed by the National Coal Board to accommodate miners brought to local collieries from declining coalfields, (ii) natural increase and (iii) an influx of commuters from the Potteries.

Biddulph is a compact town, quite distinct from the Potteries conurbation, although close to it, with easily distinguished housing areas resulting from rapid growth. The aim of this research was to ascertain whether any groups of householders distinguished by their movement characteristics existed in

Biddulph. These characteristics included movements in housing, employment, shopping and education. Such patterns of movements were correlated with the geography and recent history of the town.

Data Collection

In December 1965, some 600 households were visited within the main contiguously built-up area of Biddulph and completed questionnaires obtained from 544 households. The urban area was split up into seventeen residential units employing the combined criteria of age and type of housing. (Fig.2). Three "core" areas, the small original settlements, were identified, to which have been added five council estates, an N.C.B. estate, four estates of private development and four mixed areas, in which housing representatives of several different stages of the evolution of today's Biddulph can be found.

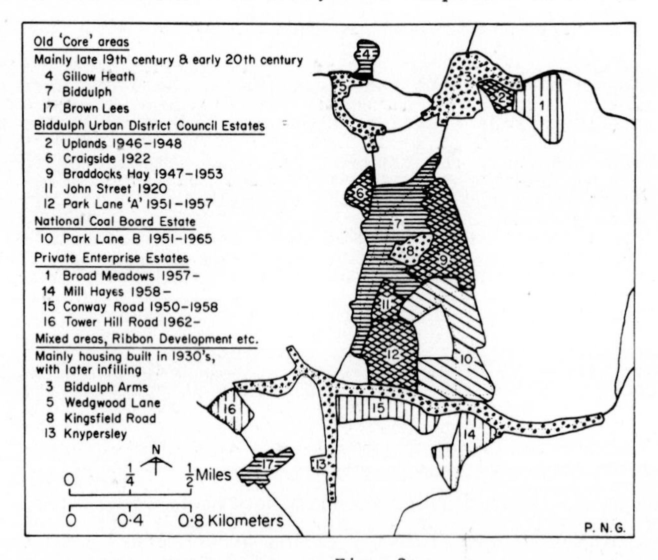

Fig. 2.

A large systematic sample of about 20 per cent was obtained from each of the seventeen residential units. Each questionnaire, with a serial number and its appropriate residential unit number, contained thirty-one questions with a total of seventy-five answers. These answers were employed as the descriptive attributes of each householder in the cluster analysis, and were either parametric or multistate non-parametric in form.

Four types of information were collected about each household, namely (i) basic information including the age and status of householder and the size of the household, (ii) residence movements relating to length of occupancy of present home, reason for moving to Biddulph and in particular to the present house, and location of previous homes, (iii) employment movements of householder: place of employment and mode of travel, previous job location and, if she worked, employment location of wife, and (iv) main shopping movements including frequency and method of travel to several shopping locations.

Application of Single Link Analysis

The methods of numerical taxonomy seemed well suited for the task of determing whether any structure existed amongst the households (the O.T.U's) in terms of their movement characteristics measured by the answers given in the questionnaire (the attributes). Computer programs for cluster analysis have been written by M.J. Shepherd and A.J. Willmott (1968)*.

The measure of "likeness" between each and every household, the similarity coefficient was calculated by treating the majority of the attributes using simple matches, in which exact matches only are considered. Certain of the attributes could not be dealt with in this manner. When numerical measures were involved in, for example, the length of time in the present home, a modified contribution, S_A to the coefficient was employed based simply on the difference between the two O.T.U's.

The similarity coefficient used here was thus a combination of the simple matching coefficient on the non-parametric attributes and the difference coefficient for the remainder.

* M.J. Shepherd and A.J. Willmott, "Cluster Analysis on the Atlas Computer", *Computer Journal*, *11*, (1968), 57-62.

The complete coefficient was standardised in the range 0-1000.

The computation of the 544 square matrix of similarity coefficients took account of the symmetry of the matrix but still involved 35 minutes of calculation on Atlas. The subsequent single-link analysis of this matrix was performed on Atlas quite quickly in less than four minutes.

The dendrogram of the Biddulph households is constructed of those clusters generated by the single-link analysis which contain at least four households. This is presented in Fig.3. Eight important branches on the dendrogram have been identified and in Fig.3, they are enclosed by dotted lines and distinguished by Roman numerals. These branches represent a hierarchial structure within the data; as movement proceeds upwards, that is as the level of similarities is raised, the finer detail of the movement patterns becomes apparent as one cluster breaks up into two or more distinguishable clusters.

Consider the large branch *I* of N.C.B. Employees. Generally speaking, these are large households, averaging over four people and living mainly on the N.C.B. Park Lane "B" estates. Virtually all are young or middle aged married men. Ninety-six per cent originate outside North Staffordshire, coming to Biddulph fairly recently from other coalfields. However the sub-branch $E \rightarrow B \rightarrow A$ is formed almost entirely of young N.C.B. householders who are employed outside Biddulph, presumably at Chatterley Whitfield Colliery. Within the branch *I* is also the sub-branch $F \rightarrow C$ containing some council estate householders who are older and some of whom work within Biddulph. Finally there is the sub-branch *M* which contains households from a variety of residential units, all of whom are employed in Biddulph.

Earlier in the paper, mention was made of an influx of commuters from the Potteries and this is represented by branches II and VII, the "New Commuters". These generally consist of people living on private estates, who have come to live in Biddulph from the Potteries. They continue to work in the Potteries and travel there by motor car. Although local and

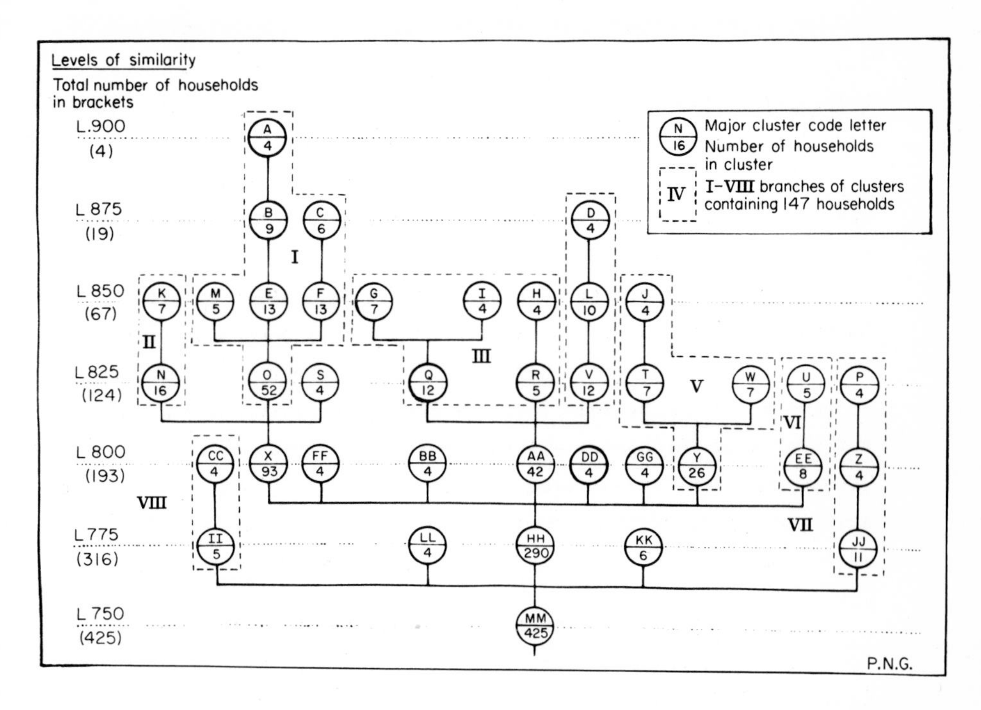

Fig. 3.

travelling shops are frequently used and most use Biddulph town centre at least weekly for shopping, about half the households shop in the Potteries at least weekly, because, they claim, there is a better choice of goods.

This branch is separate from branch VI, the "Older Commuters" which consists of mainly middle aged people living in cores, private estates and mixed areas. Most were born in Biddulph but travel by car to work in the Potteries. As with branch II, the householders of branch IV go shopping in the Potteries, half of them at least weekly.

The cluster analysis also identified the "Widows and Single Women", branch III consisting of locally born householders (sub-branch *Q*) and of those who had recently retired to Biddulph (sub-branch *R*).

The "Retired Men" appeared in branch IV while branch V was identified as the "Council Estate locals". Branch VIII consisted of the unemployed.

It can be demonstrated that a relationship exists between the branches (I to VIII) determined by household movements and the residential units determined by age and type of housing. In general it can be said that for each branch, its households are not scattered throughout all the residential units of the five groups, but are concentrated in a small number of units within not more than three of the groups.

Conclusions

What emerges from this analysis is the usefulness of Numerical Taxonomy as a method of analysis of questionnaire data in a Social Geography study such as this. The overall attribute distribution in each branch of the dendrogram represents a pattern of movement; that such a pattern is associated with a limited number of residential areas is useful because it should give some guide how best transport resources, for example, can be used in the Biddulph area.

These results show that even a relatively small urban area lacks homogeneity and that broad generalisations for the area as a whole, upon which resource allocation decisions might be made, do hide considerable internal variations. Movement patterns for the younger, more affluent and car-user sections of the community are very different from those of older, poorer, people who rely on public transport to a much greater degree.

Discussion

Q. (Paton) Was there anything established which could not have been established with conventional sociological techniques?

A.(Grimshaw) Cluster analysis gives statistical backing to the results. It is possible that other methods would have yielded similar results. Also, we wanted to see if cluster analysis could be used in our application.

(Willmott) We were anxious to try cluster analysis on a large sample. There were over 500 O.T.U's. I am not sure that all the results could have been obtained by other methods.

Q.(Parks) You started off with arbitrarily chosen regions or geographic areas. With the questionnaire data, could you by cluster analysis, derive a similar or different geographic zonation?

A.(Grimshaw) This is just what we would like to do. I would, in fact like the computer to draw me the map with the computer derived zones superimposed.

Q.(Cole) Have you tried sorting your data according to various ordering of the characteristic data, and looking at the top and bottom cases? In a historical demographic application in St. Andrews this pinpointed some interesting special cases.

A.(Grimshaw) No, we have not tried this. It might be worth trying.

Q.(Cliff) 1. Why did the authors decide on this approach (*i.e.* taxonomy) as opposed to social area analysis of Shevky and Ball which has been traditionally used to examine homogeneity of urban areas.

2. Did the authors consider, say, an analysis of variance design to test the significance of household groups identified.

A.(Grimshaw) 1. We just wanted to use cluster analysis.

2. No.

Q. (Taylor) Isn't it in a way true that if you succeed with a taxonomic technique people are going to say that it was obvious anyway?

A. (Grimshaw) Yes.

Comment (unknown) One simple way of checking on the production of "self-evident" results, which I have used myself with ecologists in the application of Association Analysis is to ask the ecologist to write down his interpretation and place his description in a sealed envelope before starting on the analysis. The interesting result is that in every case the analysis gives the information which a trained ecologist will derive from his knowledge of factors external to the data, and that in most cases, it will find something that ecologist, despite his extra knowledge, has missed. It seems to me the real value of numerical taxonomy is that it does automatically much that would otherwise have to be done by the trained man leaving him free to apply his special knowledge to extend the analysis, and that this is precisely the value that has been demonstrated in the paper we have just heard.

Q. (Jackson) 1. To what extent is your choice of classification algorithm limited by the size of machine which you use?

2. To what extent is the size of the matrix you use limited by the size of your computer memory?

A. (Willmott) 1. Paper by Willmott in Computer Journal describes clustering method whereby the rows of the similarity matrix are examined one by one. The method can be modified to deal with k-link clustering, Sørensen's method, and others but these all require some degree of back tracking of the magnetic tape which holds the matrix. Single linkage is quicker, since it requires only one examination of each row of the matrix.

A.(Willmott)2. Not at all, even on the Manchester ATLAS, which is the smallest of the ATLAS machines. The reason for the figure 2000 (O.T.U's) given earlier in the talk is that a 2000 x 2000 triangular matrix is the largest that can be stored on a magnetic tape. If more extensive data is presented, then one may use several tapes. We have not had experience of using this program on a computer with large disk storage.

MODE ANALYSIS: A GENERALIZATION OF NEAREST NEIGHBOUR WHICH REDUCES CHAINING EFFECTS

D. Wishart

Computing Laboratory, University of St. Andrews

Introduction

In 1914, the astronomer H.N. Russel plotted the temperature against luminosity of visual stars on a scatter plot, which is now known as the H-R diagram, and classified the stars into two groups which he called "giants" and "dwarfs". The diagram in figure 1 is reproduced from the H-R diagram given in Struve and Zebergs[1] which shows the dwarf star sequence as an elongated swarm from bottom right to top left, and the giant sequence as the cluster at top right. In 1963, Ward[2] proposed a method for numerical classification which was designed to optimise the error sum of squares objective function in a hierarchical fusion process. Forgey[3] applied Ward's method to the H-R diagram in 1964, and obtained the final classification into two groupings which is shown by the partition line of figure 1. The classifications of Russel and Ward clearly do not coincide, and the conclusion must surely be that, for the astronomer's purpose anyway, Ward's method failed. This paper is devoted to an examination of the reasons for that failure, a reappraisal of what a 'natural' grouping procedure should theoretically achieve, and the author's contribution of theory and method designed for taxonomic purposes. A clustering procedure, entitled Mode analysis, is developed for normal-sized data sets, and its proposed extension for large data sets is outlined.

The minimum-variance solution

The term 'minimum-variance' has been used by Forgey[3,4] to describe the basis of those methods which attempt to minimise the within-group sum of squares. In this context, any method which imposes some form of constraint on the spread, or variance, of clusters of points is included in the category. The classical

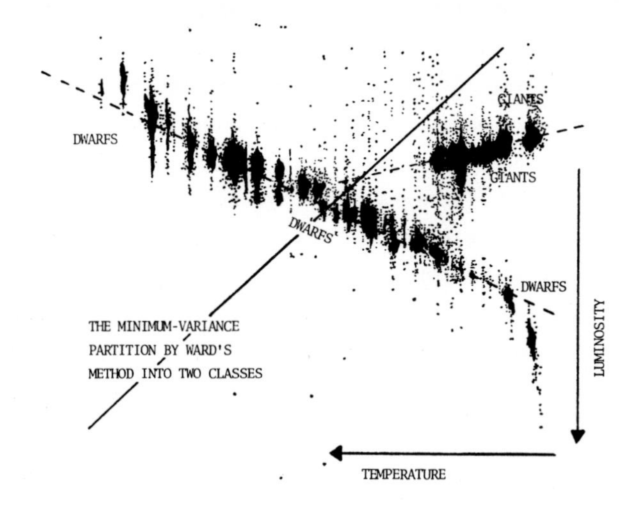

Fig. 1. The Struve-Zebergs H-R diagram for visible stars showing Russel's classification and the two class partition obtained by Ward's error sum method

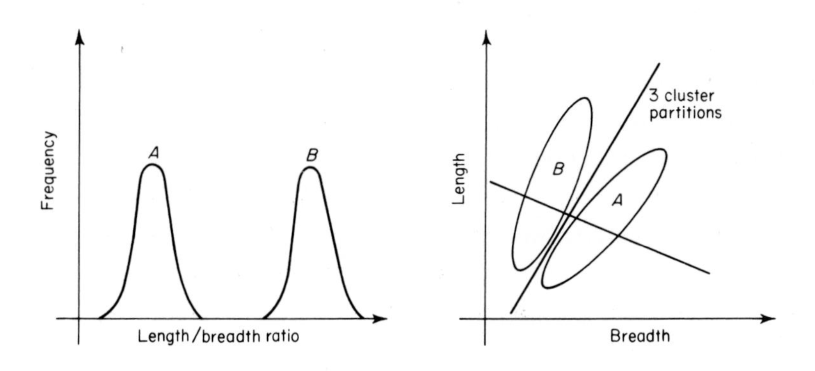

Fig. 2(a) Histogram of the length/breadth leaf ratio showing two well-defined modes associated with the artificial species A and B.

Fig. 2(b) Scatterplot of leaf length versus breadth showing elongated swarms for species A and B, and the probable division into 3 classes by a minimum-variance method.

example of this concept is exhibited by Sørensen's method[5] (1948), and the minimum-variance approach is epitomised by his statement "only one demand may justly be made on the nature of the vegetation in the limited area under investigation namely that it be homogeneous with as much approximation to that mathematical concept as nature can offer." To impose the requirement that a plant community should exhibit as near total homogeneity as is possible, that is, without any major factor of variation, is probably a perfectly valid constraint in the context of vegetation analyses. As Sørensen goes on to say "the various types of vegetation often are so insensibly merged as to form a sliding scale", and the use of a clustering method which searches for 'natural' or 'distinct' datum groupings would almost invariably fail to meet the ecologists' demands.

The consequence of such requirements has been that the major effort in the development of classification methods has been directed towards the definition of a satisfactory analysis which yields groupings that possess the minimum-variance property. The following thirteen methods are merely a representative selection from possibly, a much larger list of attempts. It is convenient, for descriptive purposes, to base the comparison of these methods solely on the use of the squared Euclidean distance, d_{ij}^2, which measures the dissimilarity between individuals i and j. Of course, it is appreciated that several of the originators of these methods specified particular similarity measures, but the trend (*e.g.* Lance and Williams[6]) has been to select the grouping procedures from these methods and generalise them for all coefficients. The use of d_{ij}^2 in this account of the methods can, therefore, be justified as simply a basis for comparison which, firstly, makes available a wide range of useful analytic reduction devices (such as vector methods), and secondly, is a widely accepted relation coefficient and liable to be adopted anyway.

Methods possessing variance constraints

The following thirteen methods possess minimum-variance constraints which are closely related, and summarized in Table 1, (the symbols used in this appraisal are defined beneath Table 1).

1. Sørensen[5] (Complete Linkage, 1948).

A group of individuals comprises a cluster provided that no two individuals have a similarity which is less than a critical user threshold r. Using d^2 this is interpreted to mean that the maximum distance between any two cluster points must not exceed the threshold, that is, the threshold defines the maximum permitted diameter of the cluster subset.

2. MacNaughton-Smith[7] (Furthest Neighbour, 1965).

Each of the N individuals is originally designated as a single-point cluster, and a hierarchy is defined by a sequence of (N-1) fusion steps for which, at each step, those two clusters having the smallest resultant diameter at union are combined. The hierarchy obtains all the possible groupings which can be derived by Sørensen's method for any threshold value.

3. Ward[2] (Error sum, 1963).

The "error sum of squares objective function" is defined as the within-group sum of squares, or the sum of the squared distances from each point to its parent cluster. The method is defined as a hierarchical process combining those two clusters whose fusion causes the least increase in the objective function at each step. This increase is easily shown (Appendix 1) to be equivalent to

$$\frac{n_P \, n_Q}{n_P + n_Q} \quad D^2_{PQ}$$

where n_P, n_Q are the sizes of clusters P and Q, and D_{PQ} the distance between their centroids. Clearly, by comparison with centroid, the method favours the union of small close clusters, and has the minimum-variance property by definition.

4. Sokal and Michener[8] (Weighted Average/Centroid, 1958).

The similarity relation between two clusters is measured by the squared distance between their centroids D^2_{PQ}, and the method can be defined as a hierarchical system (Lance and Williams[6]) which combine those two clusters, having minimum D^2_{PQ}, at each of $(N-2)$ fusion cycles. As a cluster grows in size, its centroid tends to become stabilized with the result that the similarity threshold d^2 defines a 'radius of acceptability' for the fusion of other clusters. Although the resultant spherical shape of a cluster exhibits the minimum-variance property, the centroid criterion probably imposes a weaker constraint than most of the other methods discussed here.

5. Sokal and Michener[8] (Pair group, 1958).

The relationship between a single individual i and a cluster P is defined as the average of the similarities between the individual and all the cluster elements. This can be shown (Appendix 2) to be equivalent to

$$d^2_{iP} + S^2_P$$

and, in a sequence of growth cycles, that individual for which this value is a minimum is fused to the cluster concerned. Clearly, for clusters having comparable variances, this value minimises the distance from an individual to a cluster centroid (*i.e.* reverts in a sense to centroid), while when one cluster has a large variance, then the measure favours a fusion with the smaller variance clusters (thus maintaining an upper limit on cluster variance).

6. Lance and Williams[9] (Group Average, 1966).

The concept of average linkage is developed, from pair group, for between-group relationships, the criterion of inter-group similarity being defined as the average of the similarities between all pairs of individuals, one from each group. The method is hierarchic fusion being defined for those groups which minimise the average; this can be shown (Appendix 3), when using d^2, to be equivalent to the minimisation of

$$S_P^2 + S_Q^2 + D_{PQ}^2$$

and the resultant constraint on variance is clear.

7. Bonner[10] (Method III, 1964).

A critical distance threshold r is chosen, and an individual selected at random is used as a starting point. The first cluster consists of those points which lie within a sphere of radius r about the starting point. From the remaining points, another is chosen at random to initialise the second cluster, and allocation proceeds as before. When all the points are allocated to clusters, each is re-allocated to its nearest cluster centre to form disjoint groups. The resultant clusters have a severe diameter constraint which is analogous to Sørensen's method.

8. Hyvarinen[11] (1962).

The process is identical to Bonner's except that, rather than choosing random individuals, 'typical' points are selected to initialise cluster centres, and the final clusters are defined at allocation time. An information-loss statistic is used to detect 'typical' individuals, but in the context of d^2, that point nearest the centroid of the residual set might suffice. Again, the diameter constraint is severe.

9. Ball and Hall[12] (1965).

k individuals, selected at random, initiate cluster centres, and then each of the remaining individuals is allocated to its nearest centre. The cluster centroids are computed and any two clusters P and Q are fused if D_{PQ}^2 is less than a user threshold. Also, clusters are split if the variance S_x^2 in any one dimension x exceeds another threshold S^2. Thus the variance of resultant clusters is constrained to

$$S^2 \leq MS^2$$

where M is the number of dimensions. The cluster centroids replace the original centres, and the method reallocates each datum afresh, and iterates to convergence.

The device used to split clusters epitomises the attitude of these writers that every variable should have small variation for every cluster.

10. MacQueen[13] (1966).

k random individuals are selected to initialise cluster centres. The distance from each datum to its nearest cluster centre is computed, and the point is allocated to that cluster if the distance does not exceed a threshold r; when the distance exceeds r, then the point initialises a new cluster centre. At each allocation, the new cluster centroid is computed and replaces the original cluster centre, and when the distance between two centroids becomes less than another limit, the clusters are fused. The process iterates until convergence, and final clusters have the diameter constraint $2r$.

11. Sebestyen[14] (1962).

This method resembles MacQueen's allocation algorithm with the exception that two thresholds are selected (defining spheres of radius r and R about the cluster centres, $r < R$). If the distance from a datum to its nearest centre is less than r it joins that cluster, if the distance is greater than r but less than R, the datum is set aside and allocated at a later iteration, and if the distance exceeds R then the point initialises a new cluster centre. Cluster diameters are therefore constrained to $2R$.

12. Jancey[15] (1966).

Rather than select k random individuals, Jancey selects k random points for centres and allocates each datum to its nearest cluster centre. When all the points have been allocated, the cluster centroids are computed, and the centres are moved to new positions relative to the centroids. The method then returns to reallocate and iterates to convergence. Jancey proposes an 'over-relaxation parameter', to determine the new cluster centres after reallocation, which he claims speeds up the approach to equilibrium. However, the result, at convergence, is that the final cluster centres are sited at their centroids. As such, the method does not have any marked diameter constraint, however, Jancey goes on to propose that different values of k should be tested, the optimum solution being obtained when the total within-group variance is minimised (the definition of total within-group variance is exactly the same as Ward's error sum of squares)[2].

13. Forgey[4] (1965).

In search of the ideal minimum-variance solution, Forgey adopts Ward's hierarchical process[2] to obtain a part-optimum solution for k clusters and then proceeds to reallocate cluster individuals to their nearest cluster centres. After this, he tries "sliding the partitions back and forth between each pair of centroids" in an attempt to improve the error sum of squares. The final groupings are very similar to those obtained by Jancey's method.

General comments on the minimum-variance solution.

The underlying axiom of variance constraint seems to have been developed intuitively by these writers from the idea that a resultant group of individuals should be homogeneous in relation to the total set of variables. That is, each individual should be relatively similar to every other individual in the same cluster for each variable. Expressed in geometric terms, the swarm of points which constitutes a minimum-variance cluster would be of spherical shape and should not possess any major axis of variation. Ideally, a principal components analysis of the cluster subset alone should reveal no major difference between successive latent roots indicating that the dispersion is isotropic.

	Author	Minimised Factor
1.	Sørensen: P–Q	r
2.	MacNaughton-Smith: P–Q	r
3.	Ward: P–Q	I_{PQ}
4.	Sokal and Michener (centroid): P–Q	D^2_{PQ}
5.	Sokal and Michener (pair group): E–P	$d^2_{EP}+S^2_P$
6.	Lance and Williams (group average): P–Q	$D^2_{PQ}+S^2_P+S^2_Q$
7.	Bonner: E–P	r
8.	Hyvarinen: E–P	r
9.	Ball and Hall: E–P, P–Q	d^2_{EP}, $S^2_P \leq MS^2$
10.	MacQueen: E–P, P–Q	r
11.	Sebestyen: E–P	r
12.	Jancey: E–P	d^2_{EP}, I_{PQ}
13.	Forgey: E–P	I_{PQ}, d^2_{EP}

Notes:

I_{PQ}	-	error sum of squares optimisation at fusion of clusters P and Q.
r	-	cluster subset diameter or radius constraint.
D_{PQ}	-	distance between centroids of clusters P and Q.
d_{EP}	-	distance from element E to centroid of cluster P.
S^2_P	-	variance of cluster P.
M	-	number of variables.
S^2	-	user variance threshold.
$(P$–$Q)$	-	relation between clusters P and Q.
$(E$–$P)$	-	relation between element E and cluster P.

Table 1: A comparison, based on Euclidean distance, of some clustering methods which exhibit variance constraints.

Hazards of this approach in taxonomy.

The classification of stars obtained by Russell (1914) from the H-R diagram does not possess the minimum-variance property, in fact, Russell writes "if we could put on it (the H-R diagram) some thousands of stars . . . . . we would find that the points representing them clustered principally close to two lines." These lines, dotted in figure 1, would correspond to the grouping, suggested by Russell, of stars into the two categories "giants" and "dwarfs". The sequence of dwarf stars clearly has no minimum-variance property, and indeed the idea of splitting this band into sections obviously did not occur to Russell. What impressed him was that his scatter diagram revealed two distinct modes, and it is therefore to be expected that the classification of the H-R diagram obtained by Forgey (1965), using Ward's method (1963) for optimising the error sum of squares, would never coincide with Russell's conclusion.

In their book Numerical Taxonomy,[16] Sokal and Sneath describe the traditional method by which taxa are defined as follows: "a search for characters reveals that within a subgroup *A* (of the population) certain characters appear constant, while varying in an uncorrelated manner in other subgroups. Hence a taxon is described and defined on the basis of this character complex *X*. It is assumed that this taxon is a monophyletic or 'natural' taxon". The mathematical interpretation of the constant character complex would satisfy the minimum-variance criterion since a representation of the subgroup *A* in the character space *X* would yield a small variance spherical swarm of points. Those writers who have adopted the minimum-variance approach in numerical taxonomy have extrapolated this notion to the extent that numerically derived taxa are defined by its converse. That is, for a set of characters *P*, a 'natural' taxon is a subgroup of individuals for which *P* takes constant values. Notice that the traditional taxonomist, according to Sokal and Sneath, defines the taxon from a "search . . . which reveals that . . . certain characters appear constant . . . within a subgroup *A*". This implies, as is usually the case, that while the character subspace defined by the complex *X* has a minimum-variance property, the complementary subspace, defined by those characters in *P*-*X*, certainly does not. It follows that the geometric properties of the swarm for taxon *A* in the subspace *P*-*X* would not satisfy the minimum-variance criterion, and consequently the same would be true of the total character space *P*. This can be easily verified by opening a flora at any page and selecting a species at random. Several of the characters would almost certainly be defined within wide limits indicating variation, while the distinctive characteristics of the species are probably indirect combinations of characters (for example leaf length and breadth might have wide limits of variation, when the distinctive characteristic of the leaf is, in fact, its shape).

The following particular objections to the minimum-variance approach are now outlined and discussed with liberal reference to illustrations:

The minimum-variance solution produces clusters which are-

(a) modified by changes in the character set,

(b) transformation dependent,

(c) destroyed by the introduction of non-relevant characters,

and (d) sometimes partitioned by artificial and unsatisfactory boundaries.

In order to illustrate these points, consider the two artificial species *A* and *B* which have the following characters:

	A	*B*
LEAF LENGTH	4 - 10 cm.	4 - 7 cm.
LEAF BREADTH	4 - 10 cm.	1 - 2 cm.
NUMBER OF FLOWERS	5 - 7	5 - 7

The discriminating feature present in this restricted character set is clearly the shape of the leaf; species *A* (which might well be *Nymphoides Peltata*) has spherical or orbicular leaves, while species *B* (perhaps *Myosotis Sylvatica*) possesses long, or ovate-spathulate leaves. The histogram of the ratio leaf length/breadth in figure 2(a) indicates two well defined modes corresponding to *A* and *B* such that the species would almost certainly be determined by a minimum-variance method using this single variable. On the other hand, the elongated swarms in figure 2(b), obtained from a scatterplot of length *vs.* breadth, do not possess this property, and the partition lines indicate the probable division, when three clusters are requested, which a minimum-variance method would derive. It follows that the classifications obtained from such methods depend on the original choice or manipulation of the character set.

In figure 3(a), the length/breadth ratio is plotted against the number of flowers per plant. Since the latter character is constant and non-diagnostic for both species, the swarms cluster well and satisfy minimum-variance conditions. However, the overall horizontal variance is considerably greater than the variance in the vertical direction, and consequently after the usual standardisation of variables to unit variance, the swarms would be elongated, figure 3(b), and no longer possess the minimum-variance property. Clearly, when non-diagnostic (or non-discriminate) variables are included in the character set, such transformations cause the elongation of the clusters in the subspace defined by the non-diagnostic character complex. The result might well be that the swarms are not separated in their entirety, but partitioned by arbitrary boundaries (this is particularly true of hierarchical systems where the final fusions are often inefficient).

The same situation arises when irrelevant characters are introduced. Suppose that the number of dogs within a ten-mile radius of each specimen is plotted against leaf length/breadth. If we can assume that this variable is normally distributed and independent of the other, then the resultant scatter diagram, after standardisation, would be very similar to figure 3(b). An irrelevant character, simply by its non-relevance, can be taken to be non-diagnostic, and therefore, the objections are the same as for the previous category. This possibility is less likely to arise in the context of the classification of plants where the

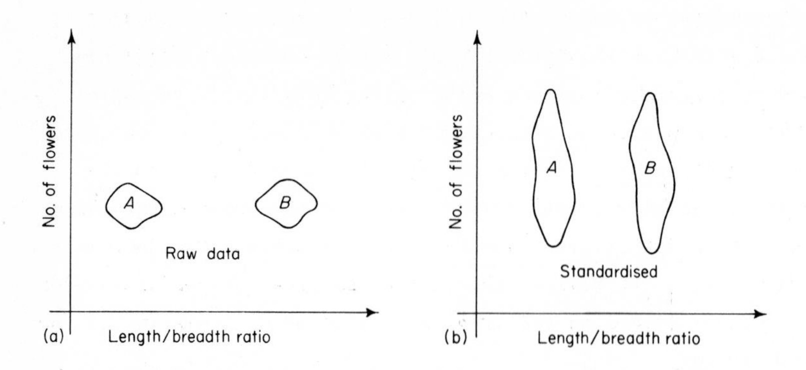

Fig. 3. Two scatterplots of 'no. of flowers' versus 'leaf length/breadth ratio'. Prior to standardisation, (a), the swarms are spherical; after standardisation, (b), the swarms are elongated and fail the minimum-variance criterion for separation.

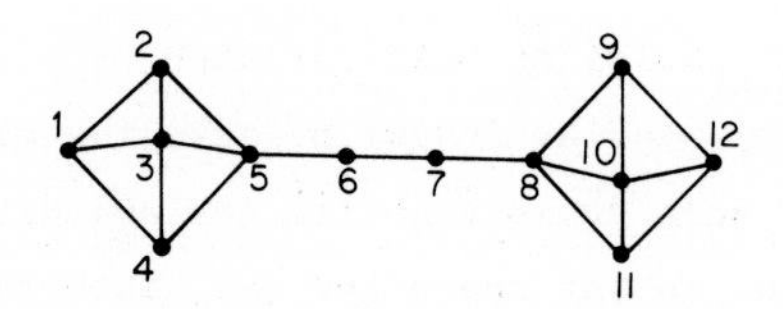

Fig. 4. An illustration of the 'chaining' effect - the lines join points which are 'similar' at some critical distance threshold. Single-linkage fails to resolve two distinct clusters due to the siting of 'noise' points 6 and 7.

characters are fairly well defined, but it would be a problem in a situation such as the classification of diseases. A patient's height might have absolutely nothing to do with his likelihood of contracting one of a group of diseases, and the inclusion of this variable, which might seem reasonable at sampling time, would result in a similar disease-swarm elongation effect.

Finally, a major objection to the partitions obtained by a minimum-variance method is that they may easily cut across a dense swarm of points (*e.g.* figures 1 and 2(b)) with the result that on either side of the partition there will be a fairly large number of individuals which are practically identical. This defeats the objective of the analysis.

The search of 'natural' classes.

Forgey[3,4] states a case for describing natural phenomena, taxa etc., in terms of disjoint data modes. He writes "when we see a frequency distribution on a continuous variable, we generally expect it to have a single mode. If there are definitely several modes, we are likely to consider our sample a mixture of several distinct types of cases . . . A scatterplot showing two distinct 'clusters' of data points suggests that the sample is a mixture of two more distinct classes of individuals . . . On the other hand, when the typical single cloud of data points is observed, it would seem arbitrary to divide the sample into any number of discrete classes." In each of the examples, figures 2 and 3, the data swarms have different shapes, but one feature in common, namely, the cluster swarms for species A and B are always separated. It would seem, therefore, that the ideal classification method for taxonomic purposes is one which can firstly tell us if there exists more than one data mode, and secondly, resolve distinct data modes regardless of their shape or variance. Classification methods which would be suitable for this purpose, that is, those which do not possess some form of variance constraint, are rare. Perhaps the most widely known is Sneath's method[17] (1957) entitled Single-linkage. A distance threshold r is determined by the user, and any two data points separated by a distance not greater than r are connected by a 'bond'. In this way, a cluster is described by a lattice of linked points, having the property that each element is 'similar' to at least one other element, while two disjoint clusters are separated by a distance which exceeds r. Another method,[18] proposed by Gengerelli (1963), turns out to be identical to single-linkage, on inspection of the linkage algorithm.

Williams, Lambert and Lance[19] (1966) have generalised the single-link algorithm to the hierarchical process 'nearest-neighbour'. Each of the n individuals initially constitutes a single element cluster, and in successive fusion cycles, those two clusters which are nearest are fused. The 'nearness' of two clusters is defined as the distance between their closest two elements, and it is easily shown that nearest-neighbour derives all the possible groupings which can be obtained from single-linkage using any threshold. This method has been severely criticised[6,19,20] for its so-called 'chaining-effects', a phenomenon which is most easily explained by reference to a diagram such as figure 4. Two distinct modes (containing points 1,2,3,4,5 and 8,9,10,11,12) are joined by the 'noise' or 'chaining' points 6 and 7 which, by their crucial siting cause the lattice of links to be extended between the modes and results in their fusion. As Forgey has pointed out, noise is a perfectly natural phenomenon of biological data where continuous variables are often normally distributed. One expects a cluster in the multidimensional space to exhibit a dense centre, or mode, which is surrounded by a cloud or noise. When attempting to classify the H-R diagram, figure 1, Forgey found that single linkage failed due to the chaining effect which occurred in the noise data that forms the 'saddle' region between the giant and dwarf star sequences. From a series of empirical trials using artificial normally-distributed data, Forgey concluded "that the method (single-linkage) performed well with very distinct clusters of any shape, but as soon as a moderate amount of noise was added the results quickly became quite erratic." Other writers have criticised the method for its chaining effect, notably Lance and Williams[6] who write "we submit that nearest-neighbour sorting should be regarded as obsolete." However, the evidence is totally empirical and few attempts seem to have been made to correct this failing.

Reducing the chaining-effect.

The obvious approach to the reduction of chaining-effects is to remove all noise data and then apply single linkage to the dense regions that are left. Forgey describes a method by which, he claims, the points are subjected to "those physical events that would occur if data points actually had mass, exerted gravitational pulls upon each other, and moved, but were not able to gather momentum." The idea is that the noise surrounding a mode would contract towards the dense centre causing the cluster to become more distinct. Unfortunately, the actual algorithm for this method does not appear to be documented, so a more detailed discussion is prohibited. Forgey does, however, concede that the method failed a test to resolve "pairs of elongated parallel clusters, even when they were made quite distinct". This may well be due to the fact that the clusters will inevitably attract each other causing them to collapse together at their mutual centre of gravity, but since the method is not known, this is pure speculation.

Sneath, however, has documented[21] a gravitation-simulation algorithm for a method developed essentially to discover the 'shape' of clusters and extrapolate a hierarchical pattern. The method seems to be theoretically satisfactory, but the complexity of its programming and user control appears to have prohibited its wide usage.

Mode Analysis - the one level test.

This is a method, introduced here for the first time, which is applied directly to the problem of removing noise data prior to single-link clustering of the denser data modes. The traditional statistical method for detecting the modes of a single continuous variable is to construct a histogram. A frequency threshold k is chosen, and the saddle regions (corresponding to class intervals which have a frequency that is less than k) are provisionally removed. The modes, if there are more than one, will now appear as groupings of the remaining class intervals (those which have a significant density) which are adjacent. Finally, the saddle regions are re-entered and associated with their nearest modes. For a scatterplot in two or more dimensions, this method is generalised to the contingency table technique whereby a grid of rectangular cells is constructed over the distribution, and the frequency of each cell is computed. Unfortunately, in order to extend the idea to M dimensions (where M is large) a contingency table of p^M cells is required when each variable is divided into p class intervals. Clearly, this would have its limitations. An alternative might be to retain only those cells which contain data points. Since each datum can only lie inside one cell, a maximum of N cells would have to be retained. The use of rectanguloid cells does, however, introduce a certain inefficiency, since to compare a data point with one cell would require M computed tests (one for each dimension). The ideal solution is to use spherical cells, since the comparison is achieved by simply measuring the distance from the point to the cell centre and comparing this with the cell radius. To ensure the maximum accuracy of mode detection when N is small, it is proposed that a cell should be located about each point, and the one-level algorithm can be stated as follows:

(a) Select a distance threshold r, and a frequency (or density) threshold k.

(b) Compute the triangular similarity matrix of all inter-point distances.

(c) Evaluate the frequency k_i of each data point, defined as the number of other points which lie within a distance r of point i (that is, those points inside a spherical cell of radius r centred at point i).

(d) Remove the 'noise' or non-dense points, those for which $k_i < k$.

(e) Cluster the remaining dense points ($k_i \geq k$) by single linkage, forming the mode nuclei.

(f) Reallocate each non-dense point to a suitable cluster according to some criterion. For the present programme, each non-dense point is included in the cluster containing its nearest dense point.

Hierarchical Mode Analysis

A major criticism of the one-level test is that two thresholds r and k must be chosen by the user. This external control can be reduced by defining a hierarchical algorithm which is based on the order in which points become dense. The method can be summarized as follows:

(a) Select the density threshold k, compute the inter-point distance matrix and the distances PD from each point to its kth. nearest point.

(b) Order the distances PD so that the smallest is first using the array KP as an index. Thus KP defines the order in which the data points become dense: point KP(1) has the smallest kth. distance PD(1) and is first to become dense when r = PD(1), point KP(2) is second at PD(2), and so on.

(c) Select distance thresholds PMIN from successive PD values, initialising a new dense point at each cycle. As the second and each subsequent dense point is introduced , the method tests the new point to determine one of three possible fusion phases:

either (i) the new point does not lie within PMIN of another dense point, in which case it initialises a new cluster mode,

(ii) the point lies within PMIN of dense points from one cluster only, and therefore the point is directly fused to that cluster,

or (iii) the point falls in the saddle region, lying within PMIN of dense points from separate clusters, and the clusters concerned are fused.

(d) Finally, a note must be kept of the nearest-neighbour distance DMIN between dense points of different clusters. When PMIN exceeds DMIN, the direct fusion of the two clusters separated by DMIN is indicated.

This algorithm is concisely represented by the flow chart in figure 5.

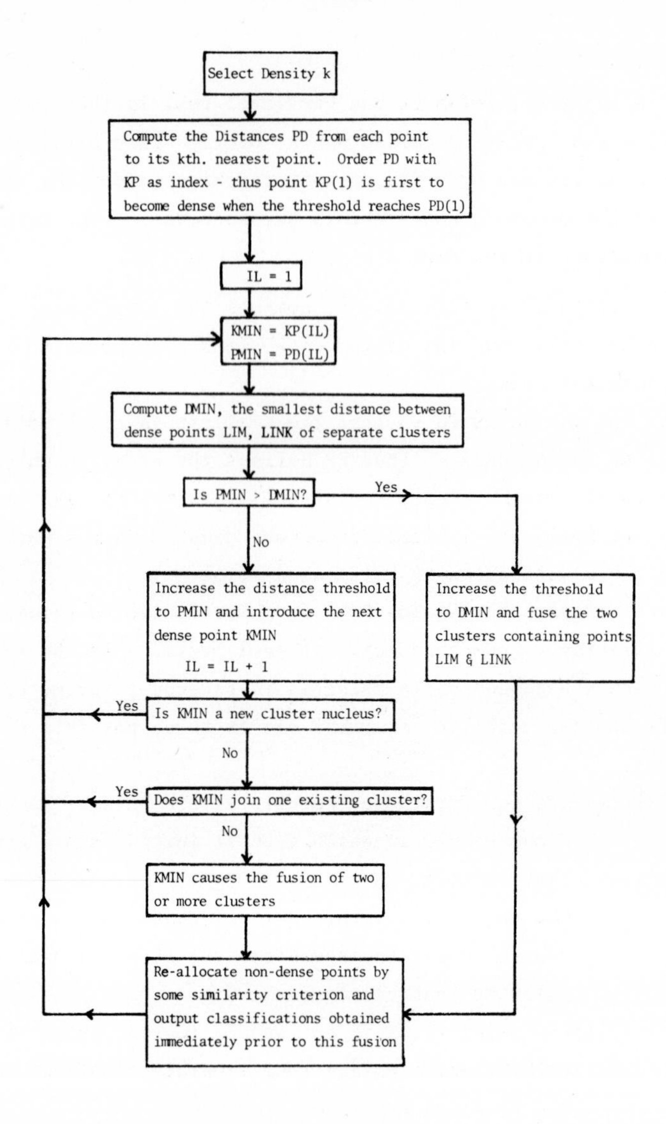

Fig. 5. Flow Chart for the Hierarchical Mode Analysis Computer Programme.

Output of classifications.

It is conceivable that, at each cycle of the algorithm, all the non-dense points could be reallocated to the cluster nuclei and the cluster groupings made available. However, this leads to a vast collection of results which are confusing, and it is therefore desirable to restrict the output in some way. The fusion of a new dense point to an existing cluster (c(ii) of the algorithm) is probably the least significant step. This can be interpreted as the growth of a mode and corresponds to an information-gain for the cluster concerned, thus the previous grouping has a lower information content and can be considered of less value. Similarly, at the introduction of a new cluster nucleus (c(i) of the algorithm), the groupings become outdated when the cluster subsequently 'grows' and increases in information-content. The really critical phases are therefore those at which existing clusters are fused (c(iii) and d), and output is restricted to those groupings which are obtained immediately before such a fusion. Two alternative levels of classification are offered to the user: the nuclei level groups only those data points (including any which are non-dense) that lie within those spheres that correspond to dense points, while at the complete classification level, each non-dense point is allocated to the cluster containing its nearest dense point. Non-dense points which lie outside dense spheres are denoted unclassifiable at the nuclei level, while, for those users who demand a best-possible fit for all their cases, the complete level of classification allocates the entire population to the cluster modes.

Unique Features of Mode Analysis

1. For the first and last cycles of the analysis only one cluster is defined. Thus, at some intermediate stage, the number of clusters reaches a maximum that can be interpreted as the widest classification which is 'natural' or 'taxonomically significant'. It is possible that an analysis will never reveal more than one cluster, indicating that the data swarm is unimodal. In a large study of several real data matrices (population sizes ranging from

30 to 350), the method never defined a grouping of more than nine clusters, and the average analysis maximum was about six.

2. The useful range of the density threshold k is about 1 to 6 depending on the population size. For large data sets ($N > 200$), empirical trials indicate that values of k in the range 3 to 5 yield practically identical results. Thus the user control is severely restricted, and seldom critical.

3. When k takes the value 1, the algorithm degenerates, by definition, to nearest neighbour, making avilable this additional method as an option for very small data sets.

4. The number of separate classifications is severely limited, by the output control, to those groupings obtained prior to cluster fusions. During the trials, the largest number of groupings obtained was 24, while the average was about 11. In one case, the method generated only six groupings for a population of size 310.

Theoretical Considerations

The histogram is usually a device for describing the shape of the probability density function of a single continuous variable. If the sample is sufficiently large, statistical tests such as chi-square can be employed to estimate the likelihood that the sampled distribution conforms to a theoretical probability density distribution, *e.g.* Normal, Poisson, etc., and when such tests prove positive we say that the sampled variable is distributed according to a particular probability function. By using contingency tables, this process is generalised to provide a means of estimating bivariate (*e.g.* binomial) and multivariate (*e.g.* multinomial) distributions. For Mode analysis, spherical cells replace the rectanguloid contingency table cells, but in all other respects the process of estimating the modes of a probability function is identical. We can say that, for large sample size, a particular spherical interval radius r and a density threshold k, those spheres which are 'dense' are sample estimators of the regions for which a complex continuous probability density function $P(u)$ takes probability values in excess of some unknown limit p. In other words, the space defined by $P(u) \geq p$ is estimated by a covering of dense spherical intervals, and if P has two or more modes at the level of probability p, then the covering will be partitioned into two or more disjoint connected subsets of points. Furthermore, if it is the case that P has more than one mode, then we can reason, by Forgey's argument, that the population is a complex mixture of several more homogeneous sub-populations which can be isolated, using Mode analysis, by partitioning the covering of dense spheres into its constituent disjoint subsets. This evaluation holds only when the sample is of sufficient size, and of course, the larger the number of dimensions (variables) which are used, the larger is the sample space and consequently the sample size must be suitably increased. At present, the programme[22] developed for Mode analysis can accommodate 400 cases, and uses each datum point to define a spherical interval. When really large populations are to be analysed, the theory would be satisfied when density spheres are chosen about a selected subset of points. In fact the traditional histogram can be thought of as a sequence of spherical (one-dimensional) density intervals selected systematically through the range of variable values. This technique would be equally valid if the intervals were chosen about actual data points selected at random. Thus, in the multi-dimensional space of a sample of size 3000, 400 points selected at random could be used to define spherical density intervals in order to locate the population modes. It is intended to proceed with this work by developing a programme along the following lines:

1. Select a subset of q key points either systematically or at random, and compute the distances PD from each key point to its kth. nearest point (from the entire population).
2. Compute the distances from each non-key point to its nearest key point.
3. Classify the key points alone using PD to define the order in which they become dense.

4. Using the complete classifications derived at each level of output, the grouping of the entire population is obtained by classifying each non-key point with the parent cluster of its nearest key point.

Finally, when particular accuracy is required, it is proposed to make a single movement of the centres of the spherical intervals in the direction of increasing density. One can imagine that spherical intervals selected at random on the fringes of two close parallel clusters might cause chaining effects. If, however, each sphere centre is moved once to the centroid of those points which it initially contains, those which lie in the saddle regions would tend to separate and become disjoint.

Appendix 1 (Ward's Method)

Suppose that the origin is at the centroid of cluster P, $\underline{d}$ is the vector joining the centroids of clusters P and Q, $\underline{u}_i$ is the vector connecting the centroid of cluster P with the ith. element of cluster P, n_P the size of cluster P, and $\underline{u}_j$, n_Q the corresponding vector and size for cluster Q. The error sum of squares for cluster P is

$$E_P = \sum_{i=1}^{n_P} \underline{u}_i^2 \text{ , and similarly } E_Q = \sum_{j=1}^{n_Q} \underline{u}_j^2$$

Since the centroid of $P \cup Q$ is located at $\dfrac{n_Q}{n_P + n_Q}\underline{d}$, the error sum of squares for the union set is given by

$$E_{PQ} = \sum_{i=1}^{n_P} \left(\underline{u}_i - \frac{n_Q}{n_P+n_Q}\underline{d}\right)^2 + \sum_{j=1}^{n_Q} \left(\underline{u}_j - \frac{n_P}{n_P+n_Q}\underline{d}\right)^2$$

$$= \sum_{i=1}^{n_P} \underline{u}_i^2 + \sum_{j=1}^{n_Q} \underline{u}_j^2 + \frac{n_P\, n_Q^2}{(n_P+n_Q)}\underline{d}^2 + \frac{n_Q\, n_P^2}{(n_P+n_Q)}\underline{d}^2$$

$$- \frac{2n_Q}{(n_P+n_Q)} \sum_{i=1}^{n_P} \underline{u}_i \cdot \underline{d} - \frac{2n_P}{(n_P + n_Q)} \sum_{j=1}^{M} \underline{u}_j \cdot \underline{d}$$

The increase in the total error sum, at the fusion PuQ is

$$I_{PQ} = E_{PQ} - E_P - E_Q$$

and since $\sum_{i=1}^{n_P} \underline{u}_i \cdot \underline{d} = \underline{d} \cdot \sum_{i=1}^{n_P} \underline{u}_i = 0$, and

similarly, $\sum_{j=1}^{n_Q} \underline{u}_j \cdot \underline{d} = 0$, the increase I_{PQ} reduces to

$$I_{PQ} = \frac{n_P \, n_Q}{n_P + n_Q} d^2$$

Appendix 2. (Pair Group: Sokal and Michener)

Let $\underline{d}$ be the vector from an outside element E to the centroid of a cluster P, $\underline{u}_i$ the vector from the centroid of P to the ith. element of P and $\underline{v}_i$ the vector connecting E to element i, then

$$\underline{v}_i = \underline{d} - \underline{u}_i$$

The average of the similarities between E and the elements of P is

$$A_{EP} = \frac{1}{n_P} \sum_{i=1}^{n_P} \underline{v}_i^2 = \frac{1}{n_P} \sum_{i=1}^{n_P} (\underline{d} - \underline{u}_i)^2$$

$$= \frac{1}{n_P} \sum_{i=1}^{n_P} \underline{d}^2 + \frac{1}{n_P} \sum_{i=1}^{n_P} \underline{u}_i^2 - \frac{2}{n_P} \sum_{i=1}^{n_P} \underline{d} \cdot \underline{u}_i$$

and since, as before $\sum_{i=1}^{n_P} \underline{u}_i \cdot \underline{d} = \underline{d} \cdot \sum_{i=1}^{n_P} \underline{u}_i = 0$,

$$A_{EP} = \underline{d}^2 + S_P^2 \quad ,$$

where S_P^2 is the variance of cluster P.

Appendix 3 (Group Average)

Suppose that the vectors joining the jth. individual of cluster Q to the centroid and ith. individual of cluster P are $\underline{d}_j$ and $\underline{v}_{ij}$ respectively, then the average of all the similarities between clusters P and Q is given by

$$A_{PQ} = \frac{1}{n_P n_Q} \sum_{j=1}^{n_Q} \sum_{i=1}^{n_P} \underline{v}_{ij}^2$$

$$= \frac{1}{n_P n_Q} \sum_{j=1}^{n_Q} (n_P \underline{d}_j^2 + n_P S_P^2) \text{ , from appendix 2,}$$

$$= \frac{1}{n_Q} \sum_{j=1}^{n_Q} \underline{d}_j^2 + S_P^2$$

Since the centroid of cluster P can be regarded as an element E, and $\frac{1}{n_Q} \sum_{j=1}^{n_Q} \underline{d}_j^2$ the average of the similarities between E and the elements of cluster Q, then, from appendix 2,

$$\frac{1}{n_Q} \sum_{j=1}^{n_Q} \underline{d}_j^2 = D_{PQ}^2 + S_Q^2$$

where D_{PQ} is the distance between the centroids of cluster P and Q.

Hence $A_{PQ} = S_P^2 + S_Q^2 + D_{PQ}^2$

References

1. Struve, O., and Zebergs, V. (1962). Astronomy of the 20th century. Macmillan, New York, p.259.

2. Ward, J.H. (1963). Hierarchical grouping to optimize an objective function, *J. Amer. Stat. Ass., 58,* p.236.

3. Forgey, E.W. (1964). Evaluation of several methods for detecting sample mixtures from different N-dimensional populations. Amer. Psychol. Ass., Los Angeles, California.

4. Forgey, E.W. (1965). Cluster analysis of multivariate data: efficiency versus interpretability of classifications. AAAS - Biometric Soc. (WNAR), Riverside, California.

5. Sørensen, T. (1948). A method of establishing groups of equal amplitude in plant sociology based on similarity of species content . . ., *Biol. Skrifter, 5,* paper 4 - (see also Sokal and Sneath[16]).

6. Lance, G.N., and Williams, W.T. (1967). A general theory of classificatory sorting strategies. I. Hierarchical systems, *Comp. J. 9*, p.373.

7. MacNaughton-Smith, P. (1965). Some statistical and other numerical techniques for classifying individuals. H.M.S.O. Home Office Research report no.6.

8. Sokal, R.R. and Michener, C.D. (1958). A statistical method for evaluating systematic relationships. *Kans. Univ. Sci. Bull. 38,* p.1409 - (see also Sokal and Sneath[16]).

9. Lance, G.N., and Williams, W.T. (1966). A generalised sorting strategy for computer classifications. *Nature, 212.* p.218 - (see also Lance and Williams[6]).

10. Bonner, R.E. (1964). On some clustering techniques. *IBM J.Res. Devlpmnt., 8*, p.22.

11. Hyvarinen, L. (1962). Classification of qualitative data. *B.I.T., 2,* p.83.

12. Ball, G.H., and Hall, D.I. (1965). ISODATA, a novel method of data analysis and pattern classification. Stanford Res. Inst., California.

13. MacQueen, J.B. (1966). Some methods for classification and analysis of multivariate observations, Western Management Sci. Inst., Univ. California. Working Paper 96.

14. Sebestyen, G.S. (1962). Pattern recognition by an adaptive process of sample set construction. IRE Trans. on Info. Theory, vol IT - 8.

15. Jancey, R.C. (1966). Multidimensional group analysis. *Aust. J. Bot.*, *14*, p.127.

16. Sokal, R.R. and Sneath, P.H.A. (1963). Principles of Numerical Taxonomy. Freeman, London.

17. Sneath, P.H.A. (1957). The application of computers to taxonomy, *J. Gen. Microbiol.*, *17*, p.201 - (see also Sokal and Sneath[16]).

18. Gengerelli, J.A. (1963). A method for detecting subgroups in a population and specifying their membership. *J. Psychol.*, *55*, p.457.

19. Williams, W.T., Lambert, J.M., and Lance, G.N., 1966. Multivariate methods in plant ecology. V. Similarity analyses and information-analyses. *J. Ecol.*, *54*, p.427.

20. Jardine, N., and Sibson, R. (1968). The construction of hierarchic and non-hierarchic classifications. *Comp. J.*, *11*, p.177.

21. Sneath, P.H.A. (1966). A method for curve seeking from scattered points, *Comp. J.*, *8*, p.383.

22. Wishart, D. (1968). A Fortran II programme for numerical classification. St. Andrews, Scotland.

Discussion

Q. (Paton) Do you consider the middle points of these two clusters to have any significance?

A. Each point, whether central or not, is merely used to estimate the probabilities obtained from a continuous density function - the centrality of the points themselves is not considered.

Q. (Jeffers) Why this obsession with the original variables? Surely by the simple technique of calculating the principal components of the original variables it becomes clear that the only 'variable' of value in the classification is the second component and that this is the best classification from the available data.

A. If one had a two-dimensional scatter diagram such as figure 2(b), and one takes principal components, the first axis will lie in the line of maximum variation and the second axis will be orthogonal to the first - but one will end up with precisely the same diagram. A transformation to principal components does not in any way change the distribution of data points it is merely a rotation of axes such that the first points along the line of greatest variation, the second is that members of the family of lines orthogonal to the first which points along the line of greatest variation, and so on. When one chooses a subset of principal component axes, the first four, say, and measures similarity distances on the basis of the scores for the first four component axes, one is merely approximating the distances which are measured in the total standardised space.

Q. (Barrs) An experience I have had in using principal components on this sort of data shows the first components to be related directly to the amount of records in any one individual. I believe Dr Jeffers has had the same problem in that the first is apparently related solely to the distribution of data and has nothing whatever to do with the actual discriminating factors.

A. Consider the following example: If one takes a series of towns and measures size of population, number of households, working people, executives etc., then in this situation one gets a principal component which corresponds to town size, or something.

Q. (Hope) While endorsing what Mr Wishart said about principal components - that you are operating in exactly the same space, I would just like to mention that one point he made about Standardisation of variables also applies to components. You can in fact standardise your components and, of course, miss them out completely; the omission of the first size component is often a handy way of getting rid of size in comparison with other variables.

A. In a situation where the first component is easily recognisable as some factor of variation in which one is not interested and, therefore, wants to eliminate, one can go back and change the character set - for example by dividing each of the other characters (in the last example) by the town population.

Comment: (Hope) This does raise unfortunate problems of spurious correlation which sometimes makes the results difficult to interpret.

A. As far as interpretation of my results is concerned I believe that one should do a principal components solution to investigate factors of variation, and then start to choose the characters which might well be based on what the results of the principal components solution are, and at the end of the analysis start looking for new characters. In a situation such as figure 2(b) where I have measured leaf length against leaf breadth, and have no significant variation on leaf length for either of the resultant clusters - I then start looking at regression coefficients between length and breadth over a cluster subset to see if there is any degree of correlation. In fact a ratio variable of length divided by breadth would be a good character to choose to distinguish these clusters.

Goronzy

1. NT-SYS system of computer programs. See F. James Rohlf, John Kishpaugh and Ron Bartcher, "Numerical Taxonomy System of Multivariate Programs", Entomology Department, The University of Kansas, Lawrence, Kansas. The system performs a great variety of multivariate analyses.
2. G.F. Bonham-Carter, "Fortran IV Program for Q-Mode Cluster Analysis of Non-Quantitative Data using IBM 7090/7094 Computers", State Geological Survey, University of Kansas, Lawrence, Kansas.

Hall

The programs are written in Manchester Autocode for an I.C.T. 1301 computer.

1. Group-forming using Heterogeneity Functions; several versions for arrays of unique items and for samples; also for two-state data, quantitative data (*e.g.* measurements and scores) and both kinds together. In a version for use in ecology, the homogeneity values for each taxon are modulated according to their relative abundances in the subsets of plots. The systems are agglomerative. Ambiguous choices for linkages (sometimes found with two-state data) are sensed and one of the alternatives may be selected by manual control. Provision is made for showing which alternative pathway gives the most compact arrangement on completion of grouping. For showing intermediateness of groups other link levels besides those chosen for fusion may be given in the output.

2. Group distinctness analysis, using the changes in heterogeneity on fusion of equal numbered sets, calculated for each property alone and for the group as a whole. Optional output of coefficients for discriminant functions for identifying unknown specimens in a critical complex.

3. Identifications for incomplete speciments known to belong to difficult groups; average heterogeneity values are found for the data of the unknown trial-linked with those of items stored as a

reference matrix in the computer. A possible match is rejected on the basis of a larger-than-specified disagreement in one of a set of chosen features. Best matches are listed in sequence of average homogeneity values and optionally the most disagreeing features may be given for each.

Ivimey-Cook

1. 1. Principal Component Analysis will read a correlation matrix preceded by its order and extract the eigenvalues and normalised eigenvectors in descending order of magnitude. Generally all eigenvalues in excess of unity are extracted, though not all may be subsequently used.

A further programme will carry out the matrix multiplication of the eigenvector matrix and the standardised data matrix.

The programme is currently available in Elliott Autocode and Elliott Algol, will be available shortly in System-4 Algol. A full specification is available.

2. Cluster Analysis

This is a modification of Dr Sneath's programme first written by G. Shad, in Leicester. It is in Elliott Algol. It carries out a cluster analysis by, optionally, pair or variable group methods, weighted or unweighted. Calculation of the revised similarity matrix is by average linkage.

3. Plot Dendrograms

This programme will read the output from the Cluster Analysis programme and will output the result in the form of a dendrogram on an incremental plotter. Currently only available in Elliott Algol.

Further information on these programmes and on certain ancillary ones can be obtained from me. Information on most of them has also appeared in the Classification Programs Newsletter.

Jackson

1. The set of programs is concerned with deriving non-hierarchical overlapping classifications for qualitative data. The programs are written in assembly code by D.M. Jackson and are available on the Titan (Atlas II) computer at the Cambridge University Mathematical Laboratory. A description of the strategy is contained in "Automatic Classification in the Ecology of the Higher Fungi" by A.F. Parker-Rhodes and D.M. Jackson contained in these Proceedings. Further details may be obtained from the authors and from the Cambridge Language Research Unit, 20 Millington Road, Cambridge.

Lerman

1. PREORDØN This one establishes a pre-order on the set "F" of all pairs of objects from a description which is provided as input. This program arranges the pairs according to decreasing similarity and edits the pre-order on F.
Here is called the subroutine ARBRE which establishes hierarchial classifications obtained by optimizing criterion of proximity between the two relations on F. ARBRE is ended up with printing a tree of classifications. This program had been written by P. Achard according to my work. It has been written in FORTRAN IV for the 3600 C.D.C. computer.
Limitations are up to 150 objects, each of them may be described by less than 1500 characters.

2. R.F. 03.
$S = (s_{ij})$ is the given matrix of similarities, where s_{ij} is the proportion of attributes possessed both by i and j. This program researches this matrix X of zeros and ones, corresponding to a partition which minimises the Euclidian distance between S and X. It is shown that the problem is equivalent to maximise one linear function whose argument is X. A process of "transfer" is used, maximising at each step the linear function; this process converges very quickly and leads to a local maximum X_1. The program starts with an arbitrary X_0.

The time of one operation from X_0 to X_1 had been 3 minutes on UNIVAC 1107 for n-500 objects into p = 20 classes.
Limitations are up to n = 500 objects - p = 20 classes.
The method is due to Mr S. Regnier and had been programmed by Mrs Renaud. These programs and others referring to different points of view especially those of W.F. La Vega and Mrs Renaud may be obtained at the following address: Centre de Calcul, M.S.H., 13 cité de Pusy, 75 - Paris 17e , France.

Orloci

1. Title of program: Ordination of relevés.

   Language: Algol 60

   Comments: Regarding technique, two alternative procedures are available. Procedure PVO is based on the use of position vectors ordination as described by Orloci (1966). Procedure COMPONENT is adopted from Grench and Thacher (1965) and Greenstadt (1960) including the Jacobi (iterative) method for finding eigen-values and vectors.

   Program description: Either of the procedures is activated by setting technique at 1 for PVO or at an integer number other than 1 for COMPONENT. The value of technique is the first number on the control card terminated by a comma.

   Regarding data input format, normally each species (character) is presented by one card or a set of cards on which species values are punched as integers, each terminated by a comma. Input can easily be formated to accommodate any type of input data.

   Regarding general strategy, apr is set at 1 for an R-analysis and at an integer number other than 1 for a Q-analysis. apr is the second number on the control card, also followed by a comma. Note that if technique is set at 1 then apr must be set at an integer number other than 1. See Orloci (1967) for explanation to terminology.

   Structure is imposed on the collection by activating procedure COEFFICIENT. This procedure computes the R- or Q-expressions of the common statistical coefficients of the scalar product type (see Orloci 1967). The identifier controlling the type of coefficient computed is cft which is set at 1, 2, or 3 for the sum of squares and products, variance - covariance, or the correlation coefficient respectively. cft is the third number on the control card, also followed by a comma.

Dimensions on input data are specified by ind (relevés or individuals) and char (species) as the fourth and fifth number on the control card each followed by a comma.

The sixth number followed by a comma on the control card specifies the number of jobs to be processed including the job in progress. Each job begins with a separate control card.

Printed output include variances of species, values in upper half of coefficient matrix including elements in principal diagonal, and ordination results. The latter include roots and vectors in the case of procedure COMPONENT, or efficiency ratios and co-ordinates on position vectors in the case of procedure PVO. If an R-analysis is followed, the proportion of root accounted for by each species is also printed.

References

Greenstadt, J. 1960. The determination of the characteristic roots of a matrix by the Jacobi method. In: Mathematical methods for digital computers. (Ed. by A. Ralston and H.S. Wilf). pp. 84-91.

Grench, R.H. and H.C. Thacher, 1965. Collected algorithms 1960-1963 from the communications of the association for computing machinery. Clearinghouse for Federal Scientific and Technical Information, U.S. Department of Commerce, Springfield, Virginia.

Orloci, L. 1966. Geometric models in ecology I. The Theory and Application of some ordination methods. *J. Ecol.* *54:* 193-215.

Orloci, L., 1967. Data centering: a review and evaluation with reference to component analysis. *Syst. Zool.*, *16*: 208-212.

Parks

1. CLUST6 is a FORTRAN IV program, presently being implemented on a CDC-6400 computer, for the classification of mixed-mode data and the drawing by the computer printer of an hierarchical dendrogram of the degree of similarity (based on a normalized simple distance function) of O.T.U's or samples. In order to eliminate bias and inadvertent weighting due to redundant and highly correlated variables, an R-type principal components analysis is performed first, and the principal component measurements on the original samples then are used to construct a distance function similarity matrix for clustering. Up to 200 variables on 1000 samples can be used. Further details and listings of the program can be obtained from: Dr James M. Parks, Director, Center for Marine and Environmental Studies, Lehigh University, Bethlehem, Pennsylvania 18015.

Ross

Programs used for classification on I.C.T. ORION at Rothamsted Experimental Station, Harpenden, Herts.

1. ORION 17. Classification Programme (CLASP). Machine Code. Reads data from paper tape or punched cards. Computes similarity matrix, minimum spanning tree, single linkage, cluster analysis, median sorting. Various forms of output.
2. ORION 14. Principle Co-ordinate Analysis. Extended Memory Autocode. Reads output from ORION 17. Also a programme to display nominated pairs of vectors. Modified to add new points.
3. ORION 52. Numerical Taxonomy Programme (NUT) E.M.A. A research programme to compare different similarity coefficients and different clustering methods.

Roux

1. ULTRAZ - A FORTRAN IV subroutine to get the subdominant ultrametric out of a given distance-matrix.
Author: M. Roux, Laboratoire de Biologie végétale, Bât. 490, Faculté des Sciences, 91 - ORSAY - France.
Storage: very small 50 Fortran IV statements. Other subroutine required: none.
Card decks, programme listing and any information may be obtained from the author.

Sackin

1. Cross-Association Versions in Elliott ALGOL and FORTRAN IV. Reads two sequences of non-numeric data (*e.g.* amino acid sequences of proteins). Slides one sequence past the other. At each overlap position counts up the number of elements which match and computes statistics indicating the goodness of matching. Also computes an overall similarity index S_L between the two sequences. Specifications of all variants of program from M.J. Sackin. Program has been published by M.J. Sackin, P.H.A. Sneath, and D.F. Merriam (1965) as Kansas Geol. Survey Special Distribution Publication no. 23. Reprints available.

2. Print Protein Pair in Specified Match Positions
Supplementary to cross-association program. Reads two protein sequences and prints them in specified overlap positions. Intended for displaying distribution of amino acid matches along position of overlap. In Elliott ALGOL. Specification available from M.J. Sackin.

Wishart

1. A programme package[22] has been developed for the IBM 1620 to classify by means of Mode analysis and eight other hierarchical techniques. The original version, in Fortran II, is currently being translated into Fortran IV for the KDF9, Hartran for Atlas, and further modifications for other systems are proposed. A feature of the package is that it is easily expanded for additional classification methods, while the basic routines possess considerable versatility of usage. A wide range of binary or numeric coefficients can be selected, and for numeric data, transformations to standard scores and principal component scores are built in, together with basic statistics such as correlations, F-tests and T-tests.

Numbers followed by an asterisk refer to pages on which the complete reference is listed.

Numbers followed by an asterisk refer to pages on which the complete reference is listed.